In praise of

BEES

Elizabeth Birchall

IN PRAISE OF
BEES

A CABINET *of* CURIOSITIES

ELIZABETH BIRCHALL

Quiller

Acknowledgements

I am deeply indebted to Polly Coles for her role as critical reader and for her confidence in the book. I cannot sufficiently thank Paul Embden, beekeeper and entomologist, for boundless access to his photographic hoard and expert appraisal of my modern scientific and practical content. I am also very grateful to Heather Leonard, beekeeper and friend, for her generous support with photographs and introduction to a demonstration meeting by the Oxfordshire Beekeepers Association at their Woodstock apiary.

I have received willing help from many staff at the Bodleian Library, Radcliffe Science Library, Natural History Museum Library, Sackler Library, and in the Print Room of the Ashmolean Museum, Oxford. Particular thanks are owed to Helen Gilio of the Bodleian Imaging Service for endless patience with obscure requests. I must also thank the following academics for help and advice on specific points in my wide-ranging book:

Dr Martin Brasier, Professor of Palaeobiology, Dept of Earth Sciences, Oxford.
Dr Paul Collins, Ashmolean Museum, Oxford
Dr Robert Johnson, Lecturer, History Faculty, Oxford.
Dr Stephen Johnston and Gemma Wright, Museum of the History of Science, Oxford.
Dr Carolyne Larrington, English Faculty, Oxford.
Dr Simon Lawson, Bodleian Library, Oxford.
Dr Claire Mellish and Dr Hilary Ketchum, Natural History Museum, London.
Katerina Nikolaidou, Vergina Archaeological Museum, Greece
Dr Heather O'Donoghue, Professor in Norse Literature, English Faculty, Oxford.
Dr Ava Oledzka, Duke Humphreys Library, Oxford
Professor Rethemiotakis, Heraklion Archaeological Museum, Crete, Greece
Dr Claire Spottiswoode, Dept of Zoology, Cambridge
Natalia Yusefovich, Kharavosk University

Other specialist help has come from:
Jennifer Davis, Archive photographer, and staff at the National Trust's Wallington Hall, Northumberland
Richard Jones, Director Emeritus of International Bee Research Association
Paula Moorhouse, The Manchester Room, Manchester City Library
Johannes Paul, Omlet Ltd, Wardington, OX17 1RR
Pat Robinson and David Bondi, Rowse Honey Ltd., Wallingford, Oxon

Thanks for images to
Herefordshire County Libraries
Rothamsted Research Ltd
Dr Naomi Saville, University College, London

Unless otherwise attributed, poems and verse passages are by the author.

Lastly, I must thank numerous friends who have patiently listened to my preoccupations and excitements over the last several years.

While exploring such varied terrain I hope I have assimilated advice willingly given but must take responsibility for any errors that may remain. I have made every reasonable effort to obtain requisite permissions to reproduce copyright material but if there have been inadvertent errors or omissions the publishers will correct these in future editions.

Elizabeth Birchall, 2014

You called them your girls, all
Hundred and sixty thousand patrolling
The flower borders, fizzing
Along the cypress corridor
Back to the hives. Their honey
Spread over our lives
Those few summers.

That blizzard winter
Of drift-blocked doors,
Already ill, you watched
From the window
While I struggled to lace
Straw duvets round the hives
But could not save lives.

First published in the UK in 2014 by
Quiller, an imprint of Quiller Publishing Ltd

British Library Cataloguing-in-Publication Data
A catalogue record for this book is available from the British Library

ISBN 978-1-84689-192-2

Printed in Hong Kong
Edited by Kirsty Ennever
Designed by Guy Callaby

*Photographs on title page and this page
courtesy of Paul Embden*

Quiller
An imprint of Quiller Publishing Ltd
Wykey House, Wykey, Shrewsbury SY4 1JA
Tel: 01939 261616 Fax: 01939 261606
E-mail: info@quillerbooks.com
Website: www.quillerpublishing.com

CONTENTS

1 : INTRODUCTION

The chiefest cause, to reade good bookes,
That moves each studious minde
Is hope, some pleasure sweet therein,
Or profit good to finde.
Now what delight can greater be
Than secrets for to knowe
Of Sacred Bees, the Muses' Birds,
All which this booke doth showe.[1]

From prehistoric cave paintings, myth and the earliest of inscriptions through to our contemporary world, it is evident that honeybees have always fascinated people, and still do. Honey would have provided important nourishment – and pleasure – until sugar arrived on the table. Now there is urgent concern about bees' fragile survival in an increasingly detrimental environment, given their cardinal importance as plant pollinators in the wild and in agriculture, the underpinnings of all life. But why did this humble insect take on such a rich freight of meanings across diverse cultures?

Cabinets of curiosities historically contained objects for enquiry or contemplation by (usually) gentlemen of leisure. I hope this cabinet of bee lore will similarly intrigue and please the reader. I have rummaged, not entirely randomly but certainly not exhaustively, in many drawers – of myth, religion, politics, moral philosophy and folklore, and also include a selection of proverbs and quotations from among many poems featuring bees as well as some of my own verse. Evidently, earlier writers were unconcerned about intellectual copyright, freely ransacking others' works, but when they disagreed invective could be fierce. I have assembled passages of literature on beekeeping from the classics; the many minds and pens that have translated Virgil have given me the pleasure of choosing variously. I draw from the exciting and excited beginnings of scientific anatomy, onward through an explosion of interest in effective hive design and the economics of beekeeping to the present day. I touch on current research programmes ranging from the major issues about bee populations to intriguing investigations into their perceptions and how their abilities might assist mankind, even in space exploration and landmine detection.

A measure of practical information has been gleaned from current authorities and my own limited hands-on experience but, while I hope to have avoided egregious errors, *In Praise of Bees* makes no claims to be an authoritative work of entomology or manual of the beekeeper's craft.

COMPANIONS IN DISCOVERY: SOME NOTABLE BEE MASTERS THROUGH THE AGES

1 Boldly borrowed from the father of English beekeeping, Charles Butler (1609) *The Feminine Monarchie*

Before moving into the thematic chapters, I offer brief details about the background and biographies of a number of prominent contributors to the bee world. Their words or ideas will be called on repeatedly in all that follows.

Aristotle (Fourth Century BC)

The great philosopher, teacher of Plato and Alexander the Great, was also a pioneering scientist and close observer of natural life who left us his *History of Animals* and *Generation of Animals*. Particularly interested in bees and keeping them in an observation hive, he made enduring discoveries, some of which were not confirmed until the nineteenth century. He was, like his successors for the next 2000 years, puzzled by bees' sex and procreative habits.

Virgil (70-19 BC)

Born near Mantua, this farmer's son became an esteemed court poet. After years of civil war during which Julius Caesar was assassinated in 44 BC, Virgil hankered for rural peace. Demobilised soldiers were resettled on his ancestral farm in the north but government compensation allowed him to buy new properties around Naples. After writing his pastoral *Eclogues*, in 30 BC he introduced his great paeon to bees in Book IV of *The Georgics*.[2]

> *Of air-born honey, gift of heaven, I now*
> *Take up the tale…*
> *Slight though the poet's theme, not slight the praise* [3]

Aged forty, he wrote of approaching death, of furling his sails and hurrying his prow to shore, leaving no time to sing of his delight in gardens.[4] In fact, he lived eleven more years and wrote his masterly *Aeneid*, about the founding of Rome. His deathbed words were 'I sang of farms and fields and men who lead'.

Charles Butler (c.1559-1647) 'The father of English beekeeping'

Born to a poor family in High Wycombe and obtaining a scholarship to Oxford, Reverend Charles Butler first worked as a teacher around Basingstoke. He also wrote – on music, grammatical matters and spelling reform. Aged about forty, he became vicar of Wootton St Lawrence, a village tucked away down Hampshire lanes though close to the town. Until he died forty-seven years later, he devoted his days to parish duties, his family, music and the study of his bees. His youngest child, Elizabeth, was his honey girl. Hive robbers and sheep rustlers, however, were to him more despicable than highwaymen.

His ground-breaking work *The Feminine Monarchie* (1609) derives much from his own beekeeping practice but, not doubting the Deity's infinite powers, he open-mindedly includes several ancient legends of bee miracles. He endorses Pliny's opinion – by giving us both nourishment and healing, the bee excels over the silkworm, which merely clothes us – it is 'to be most admired for working on behoof of men'. He admires Aristotle's close observation but scorns many later scholars who regurgitated learned ignorance, which any simple countrywoman would deride.

Most notably, Charles Butler was the first to identify the hive's monarch as a female. That the Rex must be male was an idea that several later students of the hive found difficult to abandon, not least because bees' reproductive habits were deeply mystifying for many more years.

George Wither's prefatory poem celebrates bees' exemplary virtues and the art, profit and pleasure they bring. In conclusion he writes:

2 The title derives from a word for farmer, and basically means 'the farmer's life'.

3 Virgil *Georgics* Book IV, transl. J.B. Greenough.

4 *Georgics* Book IV.

(Clockwise from above)

Fig. 1-1 : Aristotle

Photo courtesy of Bibliothèque Mazarine, Paris, and Marie-Lan Nguyen, Creative Commons licence

Fig. 1-2 : Virgil mosaic portrait

© *landesmuseum-trier@gdke.rlp.de*

Fig. 1-3 : Memorial window to Reverend Charles Butler holding his book, installed in Wootton St Lawrence church, Hampshire, in 1953

Photo by the author

Fig. 1-4 : Anthonie Van Leeuwenhoek. Mezzotint by Johannes Verkolje (c.1680)

© *Museum Boerhaave, Leiden, Netherlands*

Meethinks, there is not half that worth in Mee,
Which I have apprehended in the Bee.
Now what delight can greater be
Then secrets for to knowe,
Of sacred Bees, the Muses Birds,
All of which this booke doth shew.

Being the first substantial English book on bees and still containing much sound advice on beekeeping, *The Feminine Monarchie* became an enduring manual for centuries. 'All they doo is nothing els but woonders.'

Antonin van Leeuwenhoek (1632-1723) – wonders of the microscope

The first manageable microscope was invented in the Dutch town of Middelburg very early in the seventeenth century. But it was Leeuwenhoek, son of a Delft cloth merchant, who proved the instrument's potential. A contemporary of Vermeer and perhaps a friend, he is thought by some to have been the model for the painter's *The Geographer*.

Already a skilled lens maker, Leeuwenhoek turned his glass from examining the quality of his father's textiles to his own passionate interest in insects. Although utterly unlike our modern instruments, his simple devices had remarkably good optics. Tiny lenses ranging from x40 to x275 set in small brass plates with primitive focus and mounting mechanisms enabled him to see 'Many very little living animalcules, very prettily a-moving…'.[5] Thus he wrote of the abounding life in a specimen from his own teeth in 1683, some organisms rushing about like a pike, others spinning like tops. There was no more pleasant sight than the universe in a single drop of water.

Without any scientific training, Leeuwenhoek spent the next fifty years recording astonishing discoveries. First writing to the Royal Society in London in 1673 about the bee and the louse, his stream of letters, always in Dutch, reported a wealth of new discoveries in many fields, which led to his election as a Fellow in 1680. Always scrupulously differentiating fact from speculation, he noticed that 'all living creatures are similar in form and function'. With no established descriptive terminology, he recorded that males of all species produce animalcules, sperm. He was still writing and reporting until only a few years before he died at the age of ninety-one but many years passed before his papers were made widely available in translation.

Jan Swammerdam (1637-1680) – hard labour and piety

Leeuwenhoek's biologist contemporary Swammerdam pressed his eye hard up against one of those microscopes for twelve years. Born in Amsterdam, he was the son of an apothecary who collected natural curiosities. Having qualified as a doctor at Leiden University, Jan soon realised that his greater passion was for all other forms of life; insects and finally bees became his chief focus. His disapproving father cut off his allowance so he made ends meet by working episodically as a doctor.

God's power and divinity were made visible 'most clearly and evidently… in all created beings',[6] but he was equally devoted to Roger Bacon's precept, 'We must not feign or devise but find out… how nature operates.' Over 3000 specimens honed his expertise in dissection. He was the first to fathom insect development, finding the butterfly tightly rolled within the caterpillar's skin. The radical changes

5 www.ucmp.berkeley.edu/
history/leeuwenhoek.html

6 Jan Swammerdam (1758)
The Book of Nature, p159

from spawn to frog were incomprehensible. However, in the name of God's order he insisted this was not metamorphosis – a term then belonging to the ancient belief that insects could arise in countless random shapes from plants or putrefaction. But he did share a popular belief that all living things had existed invisibly small since the Biblical Creation. Fertilisation merely added 'a more perfect motion', perhaps life, to creatures layered (like Russian dolls) in their initial ancestors' loins.

Nature's beauties and ingenuities entranced him as much as those developmental processes. He marvels, for instance, at the feathers of bees' hairs and the way butterflies' viscous eggs lodge securely on a leaf, stuck by the mother's vulval hairs and leaving her bald. Then he focused on bees, the plainest manifestations of the invisible God.

But Swammerdam was no sentimentalist. He deduced that a full hive would contain around 50,000 cells by periodically drowning and counting the inhabitants. *The Book of Nature*, or *The history of insects* is a truly pioneering study, with revolutionary and revelatory copper plates that are still a source of wonder, such as the bee's eye (Fig. 17-2, p.222). He admires the minute architectural wonder of a honeycomb. Touching human moments and tough-mindedness abound. Since every creature starts from an egg, we cannot claim any greater dignity than a louse or a mite. Moreover, such tiny forms reveal greater marvels than do big beasts. Nature's ingenuity provides 'the strongest and most irrefragable arguments' for the eternal Godhead, showing up the atheists' case as a very weak and sorry invention of their own brain and a random conflux of atoms! His piety permits scathing dismissal of bad science; he bluntly condemns one contemporary for sometimes producing only a disordered heap of words, and others' vain theories that the net of hexagons forming a bee's eye determines why honeycombs take the form they do.

> *It would be as natural to say we should build only round houses [but] the desire of writing is so prevalent nowadays that men publish books filled only with the fancies of their brain and thus misrepresent God and his works!* [7]

He died of malaria. Like Leeuwenhoek's, his papers remained untranslated and obscure for many years but his devotion to microscopy guaranteed an enduring reputation.

Samuel Purchas (Seventeenth Century)

Unlike his parson father, who had a reputation as an unreliable chronicler of exotic travels, Samuel's clerical life in an Essex parish seems to have left few biographical traces. He was evidently both a careful observer of bees and a philosopher on the subject. His delightfully titled *Theatre of Politicall Flying Insects* first appeared in 1625, dedicated in a fulsome manner to the Earl of Warwick and setting out:

> *...the Nature, the Worth, the Work, the Wonder, and the manner of Right-ordering of the BEE... Together with Discourses Historical, and Observations Physical concerning them.*

Dipping into three centuries of Meditations, and Observations Theological and Moral, he nevertheless recalls the bookless Anthony the Hermit's riposte to a patronising philosopher – 'the world is God's Library, God manifested…' So, in the same vein, he humbly commends this study of Nature's miracles 'in the smallest and most contemptible creatures' and asks why we marvel more at elephants.

Among an array of 'poetic tributaries', Joseph Angier clearly prefers the book to its living subjects. Each word, appropriately dismissive of others' fly-blown fancies, is 'a bell from whence Mellifluous dews distil'.

Purchas acknowledges many of the Classics but rejects embroidered absurdities and scrupulously reports only what he sees. In describing bees' anatomy and habits, he forestalls the reader's potential boredom by insisting he 'cannot shut a long foot into a little shooe'.

John Thorley (1671-1759)

Yet another clergyman, Thorley was a Presbyterian minister in Chipping Norton, Oxfordshire. His splendidly titled study of bees, *Melisselogia or the Female Monarchy: An Enquiry into the Nature, Order and Government of Bees, Those Most Admirable, Instructive and Useful Insects*, first appeared in 1744. Later editions were revised by his son, who ran a business in beekeeping goods.

As usual, Thorley harks back to Classical sources and old habits of philosophising but shares the new commitment to critical observation. Incontrovertibly he proved the monarch to be female when she laid eggs in his hand.

A novel aim of his was to make beekeeping a national economic asset by improving beekeepers' skills. Eloquent in bees' defence, he advocates collecting honey without harming the providers. Why should we think it a disgrace to care for their safety and welfare, as God cares for all his creatures? Moreover, since they feed the belly and heal the body,[8] he declares:

> *They are in reality the most noble and excellent, most charming and valuable of all insects [and] deserve to be loved and defended by all.*[9]

René-Antoine Ferchault de Réaumur (1683-1757)

Another rigorous scientist, Réaumur was born to a prosperous family in La Rochelle, a few years after Swammerdam died. Educated in philosophy, law, physics and maths in various French colleges, when only twenty-four he was elected to the Académie des Sciences and wrote diverse government and academic papers but natural history was his great love.

The Ancients who held that bees were born of carcases – noble lions giving birth to kings, cows to biddable workers and a calf only to foibles [sic] – were no stupider than modern minds, but he luckily lives in a Cartesian age of reason. Even so, like Swammerdam, he supports the theory of 'preformation', every generation since Creation being layered inside the egg.

In six massive, illustrated volumes, *Notes to serve for a history of insects*, he recorded experiments and observations covering all known insects – except beetles! He examined a number of wild bee species, some making silk nests and others very pretty nests of leaf fragments, but the honeybees in his Charenton garden and at the Observatoire occupied most of his attention.

New marvels replace falsehoods, so modern works on rural economy should challenge even Virgil's admirable literature. However much we value their products and however often the hive has been seen as an object lesson in government, bees do not have us in mind as they work! Neither do we have to believe their behaviours are charitable or moral acts. Old ideas about their fastidiousness and the beekeeper's

8 Echoing Pliny and Butler.

9 John Thorley (1744) *Melisselogia*, p3.

need for purity are ill-grounded; a pomaded and powdered peruke never offends them any more than a hat and they often go near urine. They like violets, so why not violet scent? But – which particular smells do upset them?

Though a few scientists since have revealed new wonders, what a shame, he says, that Swammerdam's great work was so long lost in Dutch shadows. Nevertheless, there is still much to learn even from outside the hive – reaching into the bees' world without perturbing them. Laden bees at the doors, more crowded than a market place; fairly frequent combats; a bee struggling to carry a sister's corpse a decent distance away; young, vigorous ones seemingly killing the old and worn. If you leave them to buzz but watch and listen constantly, all this is easy to see.

Only a person born without curiosity could be uninterested in how they build their honey vases, perform their different tasks and sleep in chains and garlands. What are their social laws? A student needs to watch closely; Cassini's windowed hives are much more successful than the Ancients' horn windows or Swammerdam's paper screens. Among other discoveries, we now know the guards stand in narrowed passages and that combs are shaped by the style of hive, but much remains deeply obscure or lost among dense crowds.

Réaumur then designed a hive that allowed views from all quarters although most of his observations were from a simple glass box. His ground-breaking work on reproduction is included in Chapter Nine.

Known to friends as the 'eighteenth century Pliny', he died at the age of seventy-four after falling from a horse.

François Huber (1750-1831)

Born into a Geneva family involved in Voltaire's intellectual coterie, Huber began to lose his sight at the age of fifteen. His anonymous translator marvels that he worked through a servant, the 'truly philosophic' François Burnens.[10] Huber's wife, Marie-Aimée Lullin, also observed and assisted, a fact regrettably overlooked by the translator. In his lifetime of meticulous experimental designs and logical conclusions, 'there is evidently little room for errors of importance'. After Huber's death Burnens continued independently. Some papers were pure science while others were about bee husbandry. Notably, they finally got the bee's life history straight.

Reluctantly engaged in the servile task of translation but recognising the fundamental value of Huber's *Nouvelles Observations sur les Abeilles*, he retains the author's style, 'not elegant [but] plain and perspicuous'. The popular mind clings to historic absurdities but his Leaf Hive's glazed and hinged frames overcome obstacles that had thwarted Swammerdam and Réaumur. By examining single combs of bees at work without them 'exhibiting too formidable symptoms of displeasure',[11] he clarifies many aspects of hive life.

Maurice Polydore Marie Bernard Maeterlinck (1862-1949)

A Belgian who supported workers' rights and socialist ideals, Maeterlinck moved to France to escape his parents' and the church's disapproval of his partnership with a married woman. Though his varied literary works met with a mixed response he received a Belgian government prize and later became a Nobel laureate. His play *Pelléas et Melissande* is the basis of Debussy's opera and several other musical pieces.

The Life of the Bee,[12] published in 1901, is a meditative essay stemming from twenty years of beekeeping. He first fell in love with them in Flanders, in an old sage's

10 François Huber (1789, 2nd ed 1808 Edinburgh), *New Observations on the Natural History of Bees.* Preface.

11 Huber (1821 ed), p5.

12 He also wrote essays on ants and termites, one of which raised a serious complaint of plagiarism.

Fig. 1-5 : John Thorley Senior in his study searching for the queen among a cluster of bees, with hives and a swarm clustering on a branch outside. Shelfmark 18951 e 65, Plate 4
© The Bodleian Libraries, The University of Oxford

Fig. 1-6 : René-Antoine Ferchault de Réaumur, etching by Ambroise Tardieu
© BnF, Bibliothèque National de France

Fig. 1-7 : François Huber
Courtesy of Stéphane Fischer, photographer, and Musée d'Histoire des Sciences, Ville de Genève, Switzerland

garden where a dozen skeps were painted yellow, pink and, he asserts, the bees' favourite colour, pale blue. There he watched their humming aerial crossroads. He described his book as neither scientific treatise, nor practical guide to apiculture, nor flight of fancy but knowledgeable and loving.

A wilful but enchanting mixture of scientific truth and romanticism, the book became a little classic but some ideas strike doubtfully on modern minds. Like so many of his predecessors, he anthropomorphises and moralises. His views of the love flight and of the colony's bonds are indeed fanciful. However, alluding to Nature, God and higher or 'blind purpose', he speculates whether ideas arise in a creature's intellect or nature's mechanisms, and whether humans are in the same grip.

Karl von Frisch (1886-1982)

Awarded the Nobel Prize in 1973, this great bee scientist is most widely known for comprehending bees' dance communications. Born to an academic family in Vienna and educated there, as a child he was broadly interested in insects but focused on fish as a student. From the age of twenty-six he returned to his first love, especially bees. Most of his career was spent as a zoology professor at Munich University, researching many aspects of bee senses and behaviour for forty years.

Sometimes he writes poetically, for instance comparing the grainy image cast by bees' compound eyes to needlepoint embroidery. His portrayal of the swarm scouts' dances is vivid:

> *Just as the dances become exceedingly lively in the case of a rich harvest and become fainter and fainter with decreasing abundance of the nectar, so the scouts too dance with an intensity in proportion to the suitability of the discovery. [In due course,] something really extraordinary happens. The most vivacious of the dancers who have found a selected place gain more and more followers; these have taken a good look at it and show their approval by making propaganda for it by dancing themselves.* [13]

Some dancers are converted and others resign but the outcome is that, when the swarm flies, hundreds already know the way.

More disturbingly, his popular biology text published in the Nazi era, *You and Life*, advocates a eugenics policy that mirrors nature's ruthlessness, to prevent defectives overwhelming the 'most valuable' human stock. Worker bees' instincts are depicted in terms redolent of women's *kinder, kirche, küche* role.

Eva Crane (1912-2007)

Anyone enquiring about bees is very soon led to Eva Crane's monumental writings and, after reading her *World History of Bees and Beekeeping*, wonders what else there is to say or discover. She was a lighthouse figure towering over all things apian throughout the second half of the twentieth century.

Living to the age of ninety-five, her early academic career was in mathematics and physics followed by biomedical research. Wartime sugar shortages provoked her to acquire her first colony and another hive arrived as a wedding present. From 1940 onwards she devoted her life to the insects but, initially ambivalent, she wrote exasperatedly to her naval husband about being stung or the damn bees swarming again. However, scientific curiosity soon asserted itself. She says:

13 Frisch, Karl von (1954) *The Dancing Bees.*

Fig. 1-8 : Maurice Maeterlinck

Courtesy of US Library of Congress Bain Collection

Fig. 1-9 : Karl von Frisch

© LMU Munich

Fig. 1-10 : Dr Eva Crane with her books

© Eva Crane Trust and IBRA

It wasn't the bees I was attracted to at all. I am a scientist and I wanted to know how they worked.[14]

In 1945, litanies of questions reached husband Jim about beekeeping in China and Japan. Later her commitment continued to expand; she founded several prestigious journals and the internationally focused Bee Research Association in 1949. As time went by, bee business took over their home and the couple bought a bolt-hole in the west of Ireland. Travelling the world, sometimes by dugout canoe or dog sled, teaching and learning, her journeys are recounted in *Making a Beeline*. Ohio State University's honorary doctorate in 1985 acknowledged her own contribution and IBRA's role in promoting bee craft and hence 'the overall quality of life in many developing countries'.

After retiring in 1983, her writing continued and her collected artefacts passed to the IBRA Historical Collection in Brussels. Obituaries variously refer to her as 'queen bee', 'a modest person with a piercing curiosity' and 'an intellect that took no prisoners'.

Brother Adam (1898-1996)

As a boy Karl Kehrle moved from Germany to the Benedictine Buckfast Abbey School. From 1915 he assisted the bee master but took over in 1919 with an intensely professional and businesslike focus on all aspects of the monastery's apiary and bee products.

Heeding King Solomon's misadventures with foreign females[15] and preferring 'the

14 *Making a Beeline* (2003) Ch 2, p13.

15 E. Edwardes (1920) *The Beemaster of Warrilow*, p23.

Fig. 1-11 : Brother Adam, bee master of Buckfast Abbey
By kind permission of the Trustees of Buckfast Abbey

good old English black', traditionalists had ignored the fashion for Ligurian or Carniolan bees. However, when disease almost obliterated the Abbey's stock, they re-equipped with queens from the Austrian/Slovenian border. Then, isolated on Dartmoor, Brother Adam set about purposeful cross-breeding and established the highly sought after 'Buckfast bee', combining the Alpine's disease resistance and British hardiness.

Travelling extensively in remote central Europe and all around the Mediterranean, he scientifically catalogued pure strains of bees and evaluated their potentials as breeding stock. Such was his perseverance that at the age of ninety he was carried up Kilimanjaro in a cane chair. Besides several books on apiary management, *In Search of the Best Strains of Bees* brilliantly records those travels.

Packed with hard-nosed observation, his work comments tersely on the vile temper of the native French bee and 'some of the most horrible mongrels I have ever seen', but also lyrically describes other experiences. The Algerian desert was not honey bee country but he was delighted to see it blooming:

> *...in its full but ephemeral springtime glory — a dense carpet of desert flowers, stretching to the horizon in every direction. The air was heavy laden with the sweet scent of honey, and the traffic of insects gave the impression of a large number of swarms crossing to and fro overhead.* [16]

Brother Adam gave up beekeeping himself at the age of ninety-three but Buckfast continues its research and breeding programme.

[16] Quoted by kind permission of the Trustees of Buckfast Abbey.

2 : FROM WILD BEES TO EARLY BEEKEEPING

NATURAL HISTORY

Amber trapped the earliest known bee 100 million years ago[1] in northern Burma, a three-millimetre pollen-collector well paired with the tiny flowers then emerging. However, such early Cretaceous finds are rare and the most diverse primitive specimens come from Baltic amber aged a mere 40 to 55 million years.

The seven major species of *Apes* are a tiny subgroup of a superfamily of around 20,000 *apioides*, just a few of which have rudimentary social structures. Some, such as bumblebees, form ground nests where the queen alone survives the winter, while mason bees live alone in any convenient small cavity. Relatively recently, between 20 and 10 million years ago, social bees began storing honey but species with short tongues and no pollen baskets are poorly equipped for foraging.

AMBER

This honey icicle is warm to touch —
Cupped in the hand its smooth face invites
Intimacy. Behind, it lightly wears
A weave tracing the tree from which it leached,
And trapped within its golden pane
Are pollen grains that blew and a bee
That flew and supped on cherry sap
One summer day many million years ago.

Trees die and fall, sink slowly
To mulch and peat. Sheltered from light
And air, amber endures. A river
Bears it to the sea and there it floats
Until stranded on the tideline among
Bladderwrack, plastic, dulled glass and frayed ropes
Where beachcombers rake in hope of treasures,
Perhaps a piece that holds a tiny bee

But rarely, for bees soon acquired the art
Of powerful flight and pulled away in time.
And yet scientists found enough to chart
Their evolution from early wasps
To our familiar garden friends. This year
A fragment from the Weald's Cretaceous times
Washed up on Bexhill's beach — a wasp cocoon
Latticed with finest amber threads.

1 G.O. Poinar Jr, and B.N. Danforth (2006) 'A Fossil Bee from Early Cretaceous Burmese Amber', in Science vol. 314, no. 5799.

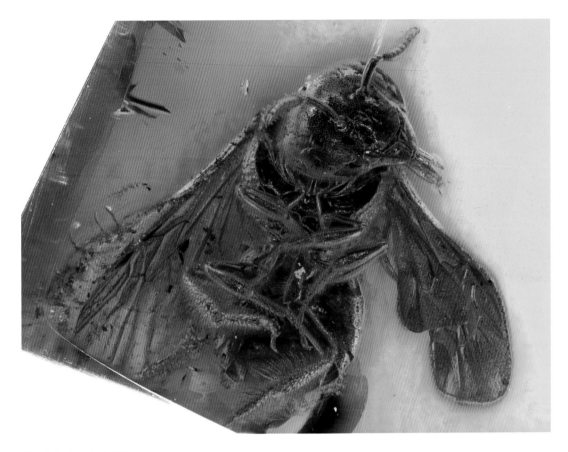

Fig. 2-1 : Bee in Baltic amber
© Anders Leth Damgaard ww.amber-inclusions.dk

The honeybee (*apis mellifera*) is the most successful and prosperous species, living in complex, self-perpetuating societies with populations up to 60,000, hanging from rock faces, filling hollow trees or gladly occupying man-made hives. Historically esteemed for their honey, even more important to the whole biosphere is their great efficiency as pollinators.

Originating in Africa and spreading from the Cape of Good Hope to the Urals, several subspecies developed into four branches: African, Oriental, North Mediterranean and West European. The European branch survived the Ice Age in Iberia and the Balkans. Experts differ on whether humans shipped them to the British Isles or whether they came via the ancient land bridge. As so often, tradition credits a holy man, seventh century St Mo Domnóc, with bringing a swarm to Ireland – but the Old Irish language suggests a much earlier arrival.

HONEY HUNTING

Cave paintings are the oldest proof of honey's importance long before mankind kept bees. Shortly after the ice receded, Spanish cave dwellers drew naked people working together with ropes and baskets to gather honey from cliff faces. Older still,

 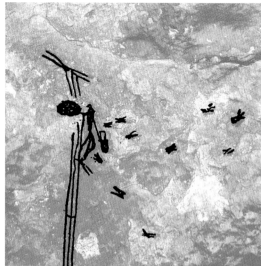

Fig. 2-2 : The honey hunter on the rock shelter at La Araña, Spain
Fig. 2-2a : Drawing by E. Hernandez-Pacheko

*Courtesy of Generalitat Valenciano. Archivo Grafico de la Conselleria de Educación,
Cultura y Deporte*

over 15,000 years ago, southern Africa's rock art shows they had learned to pacify bees with smoke. Bees and honey hunting are prominent in the 5000-year-old Indian *Rig Veda*. Seventeenth-century Samuel Purchas knew of honey trading in Angola and Benin.

These timeless techniques survive in the ancient forests of Africa, Russia and Asia. Methods of gathering from cliff nests are basically similar and all perilous, involving descents on swaying rope ladders with knives on long poles like giant chopsticks among clouds of angry bees. Combs three metres long may house 100,000 bees rippling like a Mexican wave to scare off hornets or, melting, may fall dangerously. Placatory rituals and taboos are required.

In the Naga Hills, for instance, tools are prepared in total silence and honey gathering is forbidden for three days after an animal is born. With a ceremony of wild dance, mime and singing, Sri Lankan hunters invoke their dead old women's spirits (*Maha Yakino*) to protect them and bring good fortune. The rock spirit must be appeased with sprinkled honey.

In Nepal, strict customs dictate which family inherits the job: the head descends a ladder from above, then cuts and drops the loaded combs into dangled baskets. He alone speaks to the gods and leads the preparatory rituals; sexual abstinence is required, faces are painted, garlands worn and a sacrificed sheep placates the god of the cliff. Honey may not be collected on a Wednesday, as that is said to be the bees' birthday.

Figs. 2-4, 2-5 and 2-6 show an interesting blend of the traditional and the modern, with a barefoot climber dressed in a modern bee suit, while the woman holding the banner of flowers waits with a celebratory crowd to welcome the honey baskets' descent and others break up the combs in readiness to eat them, complete with their nutritious larvae.

(Anticlockwise from top)

Fig. 2-3 : Honey hunters of Nepal
© *Éric Tourneret*

Fig. 2-4 : Nepalese village women waiting to greet the descent of the honey baskets

Fig. 2-5 : Traditional Nepalese honey hunter in modern suit

Fig. 2-6 : Eating the first fruits in Nepal
Courtesy of Heather Leonard

COLONIAL OBSERVATIONS

Until very recently it was thought maybe ninth century Norwegian or Irish explorers brought European bees to the Americas,[2] or more probably later colonists. President Jefferson remarked that honeybees were not native Americans though a species with no sting inhabits Brazil:

> *The Indians concur with us in the tradition that it was brought from Europe, but when and by whom we know not. The bees have generally extended themselves into the country a little in advance of the settlers.*[3]

Longfellow's *Song of Hiawatha* voices the same belief,

> *… people with white faces,*
> *People of the wooden vessel*
> *From the regions of the morning…*
> *Wheresoe'er they move, before them*
> *Swarms the stinging fly, the Ahmo.*
> *Swarms the bee, the honey-maker…*

But a recent fossil discovery in Nevada[4] dramatically reveals that bees were present in the Miocene era though they later became extinct.

Seeing no sign of indigenous honeybees, the first English colonists stranded and sickening in the swamps of Virginia were condemned from afar as 'unthriftie and unprofitable Drones, which live idly'. Things improved when farm stocks and hives were brought over in 1621. The Dutch came already equipped with hives for their kitchen gardens and Scandinavian forest beekeepers brought German stocks. Thus the 'white man's fly' quickly became feral and:

> *Indians with surprise found the mouldering trees of their forests suddenly teeming with ambrosial sweets, and nothing, I am told, can exceed the greedy relish with which they banquet for the first time on this unbought luxury of the wilderness.*[5]

Late in the eighteenth century, a Highland soldier called Campbell set out attended by only 'an old faithful servant, a Dog, and gun', to investigate the agricultural potential for destitute Scots to settle around the Great Lakes and Pennsylvania. The wealth of forest honey astonished him. Others have tricked bees into revealing their nest by bribing them into a handy vessel and following each after releasing them at intervals.[6,7] Campbell was fascinated by the method retold in my poem below but a later writer shrewdly commented that it is easier to rear them at home!

CAPTAIN CAMPBELL OF THE 116TH

> *I've logged plenteous wild fowl and fish,*
> *The prospering crops and stock in clearings,*
> *Think on my Bible, bounteous Promised Lands —*
> *Here honey flows from nests in maple groves*

2 Wm F. Longgood (1985) *The Queen must die, and other affairs of bees and men*, p32.

3 Thomas Jefferson *Notes on Virginia*, p199. Electronic Text Center, University of Virginia Library.

4 David A. Grimaldi and Michael S. Engel (2005) *Evolution of the Insects*.

5 Jefferson's *Notes on the State of Virginia 1785*, Electronic Text Center, University of Virginia Library.

6 John Mills (1766) *An Essay on the Management of Bees*, p154.

7 E. Edwardes (1920) *The Beemaster of Warnlow*, pp76-9.

And a man tracks bees with such artifice —
He puts up a little fire, warms a stone
And places beeswax on it. Then hard by
On another stone, he drops vermilion.

He waits. To any passing forager
The smell of wax suggests free drinks.
Red-stained, the sated creatures set course
For home. The beeline gives a compass fix,

He times the interval before they're back —
Now he can locate and mark their tree,
Then, when it suits, return to cut it down,
Bag the colony, bear it to his hive.

Saving much time and fruitless searching, people in the African savannah have a primeval partnership with a honey guide bird. In *African Sketches* quoted by Bagster,[8] the nineteenth century traveller Thomas Pringle described this 'bee-cuckoo' in action:

> *This bird, which is of a cinerous colour, and somewhat larger than the common sparrow, is well known in South Africa for its extraordinary faculty of discovering the hives or nests of the wild bees... Being extremely fond of honey, and of the bees' eggs or larvae, and at the same time unable, without assistance, to obtain access to the bee-hives, nature has supplied the Indicator with the singular instinct of calling to its aid certain other animals, and especially man himself, to enable it to attain its object...*

8 S. Bagster (1834) *The Management of Bees*, p153-4.

Fig. 2-7 : Honey Guide Bird
(*Indicator indicator*) with
Kenyan honey hunter
Courtesy of Claire Spottiswoode

Bagster himself often saw it sitting on a tree and calling *Cherr-a-cherr!* If a passer-by shows interest,

> *...it flies on before him, in short flights, from tree to tree, till it leads him to the spot where it knows a bee-hive to be concealed. It then sits still and silent till he has extracted the honeycomb...*

Contemporary scientists have delved further into this symbiotic relationship, wherein the man also signals the bird and has smoker and tools to access the nest for its precious and versatile honey while the bird enjoys the comb and valuable gleanings of honey and grubs.[10]

ARCHAIC BEEKEEPING

The craft can be dated back at least to Minoan hieroglyphs that document hives.[11] Around 2600 BC Egyptians were using beeswax figures in sorcery and honey in religious ceremonies and as taxation. A bas-relief in the Sun Temple at Abukir actually shows figures working with clay hives and honey vessels. Profiting by the flooding Nile's abundance, they were the first to migrate their bees seasonally.

Similar evidence comes from all round the Fertile Crescent, where the distinctive colouring of the ancient 'Egyptian bee' lives on. Evidence that beekeeping mattered, Hittites formulated the earliest known bee laws some 3500 years ago. Punishment was brutal, the naked and scourged hive thief being covered with honey and left to die of stings. Archaic Greece was equally stern but Cappadocia's sanctions appear comparatively benign – mere fines ranging up to six silver shekels for stealing hives.[12]

About 600 years later a Babylonian provincial governor, learning from his subjects, domesticated hives; his report[13] is slightly edited in my verse below.

> *I, Shamash-resh-usur of Suhki*
> *Have watched the Habha-people*
> *In their mountain home, collecting*
> *And preparing honey and wax.*
> *I have brought bees down*
> *To my garden of Gabbari-ibni,*
> *Now I and the gardeners understand.*

Succeeding centuries saw the craft spread, with Palestinians probably learning from their Hittite neighbours long before Jews' Egyptian captivity. While the Bible makes abundantly clear that 'lands flowing with milk and honey' were blessed, the Dead Sea Essenes provide the first evidence of Jewish beekeeping. It has been suggested that there is a connection between their name and the ancient Ephesian cult of the bee goddess, discussed in Chapter Three.

Only a few prehistoric traces of Greek beekeeping can be found – in Cretan place names such as Melikertes[14] – but bees' timeless importance is shown by the wealth of myth and ritual use of honey. Ransome suggests that the Trojan War may

10 Isack, H.A. (1999) 'The role of culture, traditions and local knowledge in co-operative honey-hunting between man and honeyguide: A case study...', in Adams, N.J. & Slotow, R.H. (eds) *Proc. 22 Int. Ornithol. Congr.*, Durban: 1351-1357. Johannesburg: BirdLife South Africa.

11 Writing was emerging around 3000 BC. Maria Gimbutas (1982) *The Goddesses and Gods of Old Europe, 6500-3500 BC*, pp184-5.

12 Ransome, pp55-6.

13 Daniel T. Potts (1997) *Mesopotamian Civilization: The Material Foundations*, p150. Chinese records of beekeeping date back 3000 years.

14 Mel being a root of bee and honey words in many languages.

have introduced them to the craft.[15] An ambiguous passage in *The Odyssey* could indicate either kept bees or opportunistic wild bees.

> *… nearby is a cave that is shaded, and pleasant,*
> *and sacred to the nymphs who are called the Nymphs of the Well-springs,*
> *Naiads. There are mixing bowls and handled jars inside it,*
> *All of stone, and there the bees deposit their honey.*[16]

However, around 750 BC when Homer's epics were recorded, the historian Hesiod mentions thatched hives.[17] Another 150 years passed and the Athenian governor Solon regulated an evidently widespread practice, decreeing that no hive be placed within 300 feet of any neighbour's colonies.[18]

Bee Husbandry and Bee Studies in Classical Greece

In the heyday of Classical Athens there were 20,000 known stocks of bees in Attica, and honey from Mount Hymettus was particularly prized.

At the same time, scientific interest in natural history was emerging. Aristotle's *History of Animals* and *Generation of Animals* are the oldest surviving studies, probably written around 343 BC. Interestingly, he gave most attention to bees; evidently a skilled beekeeper and acute observer wrote Book IX of the History, someone now known as Pseudo-Aristotle. His empirical methods and even some findings remain valid to this day. He notes their wellbeing and diseases and how they pick up their cargo of pollen:

> *… by scrabbling busily with their front feet; these they wipe off onto their middle feet, and the middle ones onto the bent part of the hind ones, then fly, clearly weighed down.*[19]

Confused about the source of honey, he nevertheless notices how returning foragers shake themselves while several others attend them. Honey is then vomited into the cells. He wrongly classified bees into three solitary and six social sorts, including wasps. Time has disproved other ideas, for instance that all make wax combs. He categorises four sorts of garden bees – a few kings in a hive, stinging bees, stingless drones and dark, flat-bellied robbers. In notably ageist and sexist misconceptions, he thinks the older bees are hairy and do the housework while the smooth young ones forage, that the hive 'leader' must necessarily be a male and that the golden ones must be lazy like 'bright and showy… women'!

Other fanciful beliefs were that bees carry a stone for ballast against winds and that sunburn darkens them. Nearly 2000 years passed before two of his suppositions were refuted: the first about a bee living for six to seven years and the second, the monarch's sex.

His openable wicker hives with suspended combs were forgotten until the modern era. Dubiously late mediaeval authority asserted that he watched a windowed hive until the bees blinded it.

15 Hilda M. Ransome, (1937) *The Sacred Bee* p76.

16 Book xiii, 103-6, Richmond Lattimore's translation.

17 Hesiod *Theogony*, line 590ff, as an incidental remark.

18 Quoted by Plutarch in his *Life of Solon*.

19 Aristotle *History of Animals* Book 9 Ch 27 para 7.

A WINDOW ON THE HIVE

Aristotle (or perhaps his Pseudo-
Namesake) applied his eyes to a pane
Of lanthorn to learn the secret lives
Within his hive. How workers tend
Their father with respect and load
The brood with food and then how, tight
Within those sealed cells, white worms
Change and grow their legs and wings.

The first to note bees' several jobs —
Cleaning the hive or building comb,
Glueing the whole with gum from trees
Or guarding the door against thieves
And predators, working the flowers
And bringing treasures home
For colleagues to unload and store.

Only one leader could be borne —
A peculiar hum precedes a swarm
Or most uncivil war breaks out
If two or more should hatch. He judged
That idle drones deserve their fate,
To see their infant cots destroyed,
Themselves be killed by commoners
As bounteous summer closes down.

ROMAN BEEKEEPING

Romans took up beekeeping later than the Greeks. Virgil's *Georgics* Book IV is our oldest surviving source but Pliny and Cicero leave tantalising hints of previous mentors: Aristomachus, who studied the insects for fifty-eight years, and Hyliscus, who retired to the desert to pursue his apian passion.

Blending Greek literature, received wisdom and practical bee craft, Virgil delights in the buzz on sunny afternoons but knows about autumn fumigation and feeding; in extremis, a long, deep sound signals the bees' urgent need of honey and healing herbs. He pays close attention to their absolute dependence on their 'king' and heroic fidelity in battle, their division of labour and hive laws. Hives should be entrusted only to a dependable gardener. In fact, his *Georgics IV* offers a complete and charming portrait of a countryman's excellent garden fit to feed a king.

Bees' legal status as wild or domestic was ambiguous until the first century AD, when Celsus' encyclopaedic *De Artibus* records they could be owned. This gave keepers rights to their harvest and liabilities if they sting. Thereafter, drawing heavily on Virgil, several handbooks on farm management appeared under the title *De Re Rustica*. Pliny, Columella and the fourth century Palladius clearly draw from Marcus Terentius Varro, the earliest surviving version.

Varro was a distinguished officer awarded the *corona navalis* for campaigning against Mediterranean pirates, and a learned man who organised a great public library. However, political upheavals disrupted his later public service career. Finally retired, he supported a 'back to the land' movement for absentee landowners and dispossessed peasants. Charm and realism mark his book's urgent purpose, to instruct his wife on farm management:

For my 80th year admonishes me to gather up my pack before I set forth from life. Had I possessed the leisure, Fundania, I should write in a more serviceable form what now I must set forth as I can, reflecting that I must hasten; for if man is a bubble as the proverb has it, all the more so is an old man. [20]

Largely conveyed through conversations between three characters and clearly indebted to Aristotle and myths, he depends extensively on the master poet Virgil. Bees' taste for the flowers of sacred mountains proves their kinship with gods and Muses; they are called their 'winged attendants' because clapping and cymbals soon gather them. [21] Like a human political state, the hive contains king, government and fellowship. Fighting only when their labours are attacked but living 'as an army, sleeping and working regularly in turn', they obey trumpet calls to peace and war. They follow their king faithfully and will carry him on their backs if necessary. Their fruit is good for both altar and table. They dislike dirt, evil odours, sweet perfumes and lazy drones.

While knowing hive glue, propolis, comes from horse chestnut buds, he mistakenly thinks pollen fastens the combs together and that wax comes from various crops. Almond trees and charlock provide valued 'triple service' — wax, honey and bee-bread.

Apiculture — hives, pasturage and forage, feeding — is a dignified and important job. Care is necessary when migrating hives in spring or splitting colonies. Stock must be healthy; striped bees are best; black ones should be killed. Wild forest bees are the best workers. Gluttonous beekeepers will suffer like a farmer who never rests a field — hungry bees provide poor harvests.

Lucius Junius Moderatus Columella, a robustly no-nonsense writer, was another ex-soldier. Born in Cadiz early in the first century AD, he farmed in Italy for many years. He despises city life, and esteems the strength and soldierly qualities of active farmers and ploughmen. Towards the end of his life he distilled his experience in *On Agriculture*. Some 1500 years later Milton recommended this comprehensive manual to students, to improve English tillage. [22]

He gets straight down to business — 'farmers are busy folk'. [23] Virgil wrote most ornately and Celsus most elegantly but tales of bees nurturing gods are of no use to a husbandman; neither are speculations about their country of origin or their manner of reproduction. The beekeeper (*mellarius*) must be clean; dirt is worse than laziness. Well-handled bees soon become tame. He supposes drones must have some use, perhaps warming the grubs, so they should not be killed. Uninterested in how honey is created, he is precise about how to harvest it. He disagrees with keeping a winter colony warm with a disembowelled fowl!

On quality, he is canny. If you cannot open the hive to check how lusty the colony is, watch the traffic and listen for the volume of buzz. He believes a bee could live ten years but diseases often overtake them, so new swarms each spring are essential. Always check a swarm's quality — bad bees take as much labour as good ones. Local

20 Varro, *De Re Rustica* Book I, opening lines.

21 An evident reference to Dionysus/Bacchus and an interesting forerunner of 'tanging'. Myths are discussed in Chapter 3 and tanging in Chapter 8.

22 From Introduction to Vol I of Columella, by Harrison Boyd Ash.

23 Columella *De Re Rustica*. Book IX deals with bees.

bees are best. Those brought from afar should be fetched at night on a man's shoulders, not jarred over rough roads on a wagon. They must be confined for twenty-four hours, then watched in the morning to see whether they sally forth in a body contemplating escape.

When brood falls off, the old king should not be killed because 'the… older bees, who form a kind of senate, do not think to obey the juniors'. First to mention giving a weak colony a new leader from another hive, he also knows how and when to kill excess young queens – skills subsequent beekeepers lost.

Some of his advice now sounds bizarre. For instance, if bees are too busy foraging to produce offspring, they should be shut in every third day to get on with laying! When spring cleaning hives, the beekeeper must have no sex the day before, must wash and not be drunk or smell of pickled fish or garlic. More lasting principles are to take your last honey at the equinox so that the bees can build their winter stores. Then give them winter feed (of mashed figs, must or raisin wine), compact the living space, seal gaps and thatch the hives against winter.

Pliny the Elder,[24] who farmed on the banks of the Po, was another keen beekeeper. His mind was so quick that he was said to need two attendants while in his bath or carriage, one reading to him while the other jotted down his magpie gleanings.

'Lantern horn' and 'transparent stone' hive windows enabled Pliny to learn new facts. His close account of development from egg to imago errs only on the timing, but he believes the king was never a grub but is born winged, walking upright on straight legs and honey-coloured like the finest flowers. Meanwhile, worker bees brood like hens.

He embarks his hives upriver to new crops and, though moving daily, his bees always find their way back. When the ship's waterline reaches a certain level (a forerunner of the Plimsoll line!), he can return home to harvest the honey. He leaves a tenth for the bees on Vulcan's Day, 23rd August, but draws off his last portion when the vintage ends around the ides of November (13th), leaving the larger part for their winter feed.

Like Virgil, Pliny times bees' seasons by the stars. Hibernation lasts from the winter solstice to the rising of Arcturus, the red giant in the constellation of Boötes, but it is still too cold for them to emerge. Their working season used to extend from the Pleiades' rising on 23rd March to their setting on 11th November but now he says they wait for beans to flower! A bee rouses them with a trumpet-like double or triple boom and at night he:

> … *flies round with the same blast…, as though bidding them take their rest, after the fashion of a camp.* [25]

On fine days they are busy, their shoulders weighted with a stone against strong winds. If caught out by nightfall, they lie on their backs to keep dew off their wings! After work is done and their offspring are hatched, their other duties sound like pleasures – ranging abroad, flying high, wheeling, coming home to eat.

He is unsure whether stinging kills them. Cow-dung smokers are good for killing hive pests because bees are related to oxen by birth! They delight in 'clapping of hands and tinkling bells'; this gathers them home. Other curious fancies are that a woodpecker's beak in your pocket prevents stings and dust from a serpent's track will drive them home. He doubts whether a dead hive can be revived by wintering

24 Pliny *The Natural History of Animals*, mostly Book XI, various chapters.

25 Pliny, Book XI, Ch 10, para 10.

it indoors, or warming it in spring sun and the hot ashes of a fig tree.

Lastly, Palladius. It was perhaps an eccentric undertaking for Victorian Mark Liddell to render Latin into Middle English verse but the pleasing result justifies the moderate effort involved in understanding it.

> *The keper pure and chast and with him ofte,*
> *His hyvys havyng redy forto take*
> *His swarmys yonge, and sette hem feire olofte,*
> *The smelle if donge and crabbis brent, aslake*
> *Awey from hem; and placis that wole make*
> *A voys ayeyn, as ofte as me wole calle,*
> *Is nought for hem; eek nought is titymalle.*[26]

Geoponica collated many previous books for Emperor Constantine VII in 950 AD.[27] Sketchy on practice, it retails Democritus's ancient tip on destroying drones. These supposedly redundant creatures, insatiably thirsty from gorging on honey, can be fooled if hives are sprinkled inside with water in the evening. Drinking next daybreak, they are easy to capture and kill.

From Forest Beekeeping to Domestic Apiaries in Northern Europe

Hive beekeeping developed centuries later in northern countries, gradually evolving from raiding nests in hollow trees through natural holes to cutting and fitting doors into the trunks. When landowners objected to this damage, people hung up hollow logs as hives and later realised the convenience of bringing these close to home. Thus the notion of the domestic apiary developed.

Ancient laws governed owners' rights and taxation; for instance, abbey servants or bee coerls, who were with swineherds the lowest ranks of freemen,[28] were licensed to own bees or collect honey and swarms from trees bearing their mark. From early mediaeval times, they formed dignified guilds with appointed officials. In Slav lands and Germany these Zeidler policed the forests.

In Russia the penalty for obliterating a hive mark was severe – the price of fifteen cows, six horses or 120 sheep! Irish Brehon Laws and the *Bechbretha* also spell out obligations and penalties: for, instance, a man would be punished if his hens ate a neighbour's bees. (That the bees themselves could be prosecuted is explored in Chapter Fifteen.) Brehon stipulations of weights and measures for honey vessels suggest cattle were small or beekeepers very strong!

> *A milch-cow you can lift to your knee;*
> *A heifer to your waist;*
> *A smaller heifer to your shoulder;*
> *A dairt above your head!*[29]

26 Liddell (1896) *Palladius – De Re Rustico*, p70. Virgil warned against dung and burnt crabs. A 'voys ayeyn' (again) means an echo; titymalle is a plant of the spurge family. I have transposed the ME usage of 'u' and 'v' into modern usage for easier reading but, believing the rest fairly transparent, have left other spellings for their original delight.

27 Gleanings from H.M. Fraser (1931) *Beekeeping in Antiquity.*

28 Henry III charters, 1217.

29 Quoted by Ransome, p194.

(Above)

Fig. 2-8 : Forest beekeepers and Zeidler,
etching from J. G.Krünitz (1783) *Oekonomisch-
technologische Encyclopädie*. Staatlische Museen
© *bpk, Berlin*

(Opposite)

Fig. 2-9 : A farmyard with beehives in February, by les
frères de Limbourg, from *Les Très Riches Heures du Duc
de Berry*. C15. Musée Condé, Chantilly
© *RMN-Grand Palais (domaine de Chantilly) / René-Gabriel Ojéda*

MEDIAEVAL AND MONASTIC BEEKEEPING

Books such as *Les Très Riches Heures du Duc de Berry* show that lay-people kept bees and local archives would produce extensive records. One such reports in 1350 the appearance of a single hive at Merton College's manorial farm in Cuxham, Oxfordshire. Another is the aggrieved fifteenth century Welshman, Ieuwan Gethin, who makes an impassioned plea to his overlord for justice:

> *Splendid Sion ap Rhys, gentle, who breaks spears:*
> *I am making a serious complaint*
> *That you, in spite of the agreement between us,*
> *Are destroying my bees' nests.*
> *By God, your servants...*
> *They are stealing a portion of the honey...*
> *By God and Jesus, forbid Llywelyn,*
> *The dark man, your servant...*
> *The bee–hunter, early in the morning*
> *Is like a fly, searching the trees*
> *With his little axe...*
> *A hundred gallons*
> *Will have been stolen by these locusts.* [30]

30 From an unpublished fifteenth century Welsh original, transl. Dr W. Linneard and Dr A.E. Williams, and included in Crane's *World History of Beekeeping*, p114.

Fig. 2-10 : Pieter Brueghel the Elder. The Beekeepers and the Birdnester, c.1568. Staatlische Museen, Berlin

© *Jörg P. Anders, photographer, and bpk, Berlin*

Receiving short shrift, finally he appeals to the Sherriff, 'Sir William, with his clean faultless sword', who intervened and insisted poor Ieuwan be compensated.

Monasteries' need for candle wax involved them heavily in beekeeping but, despite being centres of mediaeval scholarship, they made little or no advance in bee knowledge. Two thirteenth century friars, the Dominican Thomas de Cantimpré and the Franciscan Bartolomaeus Anglicus, compiled all-embracing encyclopaedias. Cantimpré presented bees as models of Christian chastity and good community while Bartolomaeus broadened out with ideas from the Classics and also wrote another major work, *Of the Properties of Things*, which roamed widely within contemporary natural sciences. Cantimpré remained an authority until Renaissance and Reformation views challenged the mediaeval mindset. In the later sixteenth century, interest in empirical observation resurfaced.

The German Nickel Jakob overthrew the belief that the hive monarch was born fully formed, having watched workers raising her from a larva. Luis Mendez, a Spaniard, also established that she is the sole mother of the hive. These were early bricks in the bridge from the Classical era to the early modern age, soon to be followed by Charles Butler's *The Feminine Monarchie* in 1609[31] and the revelations offered by the first microscopes.

Our paths must now diverge, to view rapid and overlapping developments in bee craft and science. But first we will make a broad detour through a diverse landscape of myth, religion and folklore.

31 Despite Butler's uncompromising title, many more years passed before female supremacy could be absorbed. Bees' gender roles and sexuality remained indigestible mysteries for further centuries – see separate chapter.

3 : MESSENGERS OF THE GODS

PORTIONS OF ETHEREAL THOUGHT ENDUED WITH PARTICLES OF HEAVENLY FIRES[1]

So many cultures have venerated bees and adopted them to endorse the sanctity and authority of their kings and priests. With their intricate combs secreting golden essence, they have been seen as 'images of the miraculous interconnectedness of life'.[2]

Interest in the idea of a primordial Mother Goddess emerged in the twentieth century, a lost, possibly matriarchal society to challenge Judaeo-Christian and Islamic male monotheistic orthodoxy. James Frazer's *The Golden Bough* (1890) provided an archetype for psychological theories and then for feminist theology, deeply affecting mainstream archaeologists. Discoveries, particularly Mellaart's work on numerous figurines at Çatalhöyük, Turkey, were interpreted in this light and attracted women scholars. This archaeological furrow in South Eastern Europe and the Black Sea region was cultivated by Gimbutas, arguing 'a flourishing group of Goddess worshipping cultures' that achieved a peaceful Golden Age of Old European civilisation by 4500 BC. Her views underlie later works and cultic speculations, which argue a widespread 'goddess' belief enduring from pre-literate societies through to Classical Greece.[3]

Assertions that this primeval goddess was the direct ancestor of later belief in a 'bee goddess' intrigued me. However, scholarly debate has been vigorous. Essays edited by Goodison and Morris,[4] while not denying the metaphoric value of myth, contest any such monolithic notion. Arguing the staggering diversity of archaic beliefs across Europe and the Middle Eastern world, they warn against seeing ambiguous artefacts from a silent polytheistic past with modern western eyes.

I cannot delve deeply into that debate but the yearning for a nurturing bond with the earth and cosmos, expressed in modern myth-making about the role of bees, is interesting in its own right. Mythic narratives and religious practices provide many people with moral and political metaphors for daily life.

PART ONE

In Prehistory

Cave paintings, such as the honey hunters in Chapter Two, are generally thought to invoke supernatural assistance in the tasks of daily life.

> *Millennia before anyone had framed*
> *Linear–A, cuneiform or hieroglyphs,*
> *Votive images were painted in the depths of caves —*
> *The hunt for deer, the bison and the auroch,*
> *Even a skinny naked human climbing*
> *A flimsy ladder, clutching basket and pole,*
> *Beset by storms of bees.*

1 Virgil's *Georgics*, transl. Dryden, p115.

2 Anne Baring and Jules Cashford (1991) *The Myth of the Goddess: Evolution of an Image*, p73.

3 Marija Gimbutas (1982) *The Goddesses and Gods of Old Europe, 7000-3500 BC*; Dexter and Jones-Bley (eds, 1997) 'The fall and transformation of Old Europe: Recapitulation 1993' pp351-372 in *The Kurgan Culture and the Indo-Europeanization of Europe: Selected articles from 1952 to 1953 by Marija Gimbutas*; Baring and Cashford; www.mothergoddess.com; www.thebeegoddess.com; Andrew Gough's Arcadia 'The Bee' website.

4 Lucy Goodison and Christine Morris (1998) *Ancient Goddesses: the Myths and the Evidence*.

Many Neolithic figurines of exaggerated feminine dimensions, dating from around 5700 BC, have been found across a wide swathe of 'Old Europe' around the Black Sea and Eastern Mediterranean. Scholars argue whether they are votive figures of the Great Mother or simply symbols of fertility or of a matriarchal society. [5, 6] Or are they simply toys? Male figures and highly abstracted human shapes have also been found.

There is no evidence of a contemporary cult of bee goddesses despite some modern beliefs. Goddess cultists have very questionably attributed a pottery figure to 8000 BC and to the Hacilar site in Turkey, surmising that she is the primal goddess's manifestation as a queen bee supporting honey-streaming breasts. [7] However, this anachronistic date and a disturbing trade in modern fakes seriously muddy the waters. [8] The cult also questionably identifies figures on a gold seal ring from the Isopata grave near Knossos in Minoan Crete as bee priestesses venerating their goddess.

Fig. 3-1 : The gold seal ring from a grave at Isopata, Crete, portraying figures with insect heads and hands, c.1500 BC
© *Archaeological Museum of Heraklion-Hellenic Ministry of Culture and Sports-Tap Service*

5 Margaret Ehrenberg (1989) *Women in Prehistory*, p73.

6 Gimbutas.

7 *The Goddesses.com* website. My efforts to obtain a printable resolution of this image unfortunately failed.

8 Dr Paul Collins, Ashmolean Museum, Oxford, suggests this is a fake and knows of 'no evidence to connect such figures with bees'. (Personal communication.)

9 Baring and Cashford, p73.

10 Gimbutas, p185.

11 Paul Rehak 'The Isopata Ring and the Question of Narrative in Neopalatial Glyptic' From: kuscholarworks. ku.edu/dspace/ bitstream/1808/8364/1/ Rehak_Isopata.pdf

12 The oldest known, Fourth Millennium BC. Gimbutas, p182.

Maybe genuine archaic figures do link a primeval Great Mother to Artemis, picking up bee associations en route. [9] The evidence is clear at Ephesus, says Ehrenburg, but wider claims are greatly exaggerated. The Isopata figures have insect hands and ambiguous heads; bees and bee nymphs, Melissae, are prominent in Cretan myth but butterflies were also symbolically important. [10] Paul Rehak argues that they depict contemporary young women. [11]

Symbolic threads do, however, interweave. A female figure is often drawn on a bull's head, as in a stylised bone carving from Bilcze Zlote, Ukraine, [12] and bulls appear in many sacred contexts. The shared 'horned' appearance of crescent moons, bulls and bees' antennae hold deeply embedded associations with death and rebirth in diverse cultures, although Gimbutas says culture and belief systems in the Black Sea region at that time were only local. Debatably,[13] she identified Bronze Age remains at Apasas,[14] attributed to warlike Amazons of the southern steppes, as sanctuaries of the Mother Goddess. Nearby but much later bee priestesses served Artemis.[15]

There was a widespread and long-lasting conviction that ox carcases gave birth to bees.[16] Hittites believed it was a bee that saved creation. In variant myths, bees were the only creatures to escape Eden before the Fall or they left with Adam and Eve, expelled for corruption and doomed like Man to redeem themselves by work.

In Rhodes Museum a necklace comprised of small gold plaques of a bee goddess long antedates Classical Greece's Artemis, seeming an unambiguous link with those earlier beliefs. Later rituals show evidence of tributes gratefully offered for her bounty, as my verse records.

> *We lay our harvest best —*
> *Our grapes, honey, olives, figs —*
> *Before you, Great Mother Goddess*
> *But beware, anyone who dares*
> *To break an oath we will press*
> *Like honeycomb and molten wax.*

The bull-born Goddess of Transformation was familiar to Ovid and Virgil. In the third century AD, the Greek philosopher Porphyry links her and the bee-begetting Moon, bull-like during its ascension,[17] to Artemis and to the harvest goddess Demeter and her priestesses Melissae. Moreover, 'souls that pass to the earth are bull-begotten'. Elsewhere we will see that souls depart the earth as bees.

Ideas of fertility seem deeply woven and those female figurines surely express fecundity but the site at Apasas became the home of an insistently virginal Artemis served by celibate bee priestesses and eunuch priests. In the Graeco-Roman world of the Gospel writers this seems to metamorphose into the figure of the Blessed Virgin Mary but before picking up those themes, let us take a look at the place of bees in the earliest writings.

In the Ancient World of Written Meanings

Certain contemporary scholars assume a primordial supreme goddess in prehistoric Mesopotamia,[18] Inanna-Ishtar, goddess of Above and Below, deposed by the male god Marduk in the era of rebuilt Babylon. Furthermore, they link her with the Minoan civilisation's Dionysian rituals, when bulls were sacrificed, mead was drunk and, in Virgil's words, 'They clash the cymbals of the Great Earth Mother'.[19] But Westenholz persuasively argues this linkage is a modern myth, imposed by western monotheistic and feminist perspectives on a society with a pantheon of both genders.[20] There is no evidence of such a supreme goddess in Mesopotamia's historic period, i.e. after 3000 BC. Again, modern cultists have claimed a 'Sumerian stele' represents a bee goddess and her devotees,[21] but their image more resembles a much later, Neo-

13 Ruth Tringham and Margaret Conkey 'Rethinking Figurines', in Goodison and Morris (1998).

14 Sometimes this name is linked to the root of the word *apis*, bee.

15 See later in this chapter.

16 Discussed in Chapter Nine.

17 Hilda M. Ransome (1937) *The Sacred Bee*, p182.

18 Baring and Cashford (1991); Iris Furlong 'The Mythology of the Ancient Near East', in Carolyne Larrington, ed. (1992) *The Feminist Companion to Mythology*, p8.

19 Virgil *Georgics* Book IV, line 63.

20 Joan G. Westenholz, 'Goddesses of the Ancient Near East 3000-1000 BC', pp63-81, in Goodison and Morris.

21 Andrew Gough's *Arcadia* website, 'The Bee'.

Assyrian document seal and 'is unusual and may have been recut, and nothing in the design references a bee'.[22]

Bees or no, honey certainly was prominent among votive offerings in rebuilt Babylon and this prayer had sacerdotal power:

> *May the honey be pure for me; may it be bright for me.*
> *Let the evil tongue stand aside.*[23]

The Pharaohs' Egypt undoubtedly linked bees to the gods. Below is my free adaptation from the Opening of the Mouth ceremony.

> *Ra, you framed the earth and sea*
> *And flood the Nile to bring us life.*
> *Your tears spring forth bees to labour*
> *Among every kind of tree and flower,*
> *Supplying us with wax and honey.*
> *For it is sweet to your heart*
> *And shall never depart.*
>
> *The sun rises and we must praise —*
> *Hail Amon-Ra with honey,*
> *The Eye of Horus, the Sweet One,*
> *The tear from the Eye of Ra,*
> *The Lord of Offerings.*[24]

Bee and reed together mark the union between the Upper and Lower Nile kingdoms under King Menes. Ransome interprets the numerous bee cartouches as sacred symbols of kingship but Fraser considers they simply reflect honey's economic importance; the civil service had long included an official Sealer of the Honey.[25]

Bees, honey and mead permeate Indian religious practice; people wash their household gods in honey and Andaman Island elders use it when ceremonially feeding and bathing an initiate. The Sanskrit word *madhu* is obviously a root of several divine attributes (and its variants are evident in many Aryan languages); the Vedas' secret essence was the *madhu-vidya* or honey doctrine.[26, 27] Siva is generally characterised as the destroyer but his suaver form is Madheri with a symbolic bee. An ethereal blue bee on a lotus flower is one expression of Hindus' supreme deity Vishnu, from whose footstep springs mead; incarnated as Krishna, he wears the same blue bee.

In Hindu mythology, the insects are divine assistants to humans' earthly life. Strikingly reminiscent of Apollo, the Asvins ride the sun's chariot but their whole equipage is significantly different and laden with bee imagery.

> *All is resurrected by the sun*
> *Pouring honey, life's sap, everywhere,*
>
> *The Asvins, twin horsemen, lords of light,*
> *Drive their chariot from dawn to dark,*
> *Honey-laden and full of healing.*

22 Dr Paul Collins, Ashmolean Museum. (Personal communication.)

23 Baring and Cashford.

24 Poem freely adapted from Ransome.

25 HM Fraser (1931) *Beekeeping in Antiquity*, p4.

26 David L. Spess (2000) *Soma: the Divine Hallucinogen*.

27 The Rig Veda, 2-3000 BC.

Their white horses quench their thirst
With mead. Their whip Madhukasa
Sprinkles sweetness, food and strength
On everyone they pass.

Our final home is a meadow
With a spring of mead. [28]

A second myth tells of a conflict between the demon Arun and the nature spirits. When Arun is near to winning, the goddess Parmeshwara Deva is implored to save the nature spirits. Presenting herself in bee form, as Brahmari Devi she acquired the title of The Protector, blackening the whole earth and sky with her ferocious and victorious allies.

Fig. 3-2 : The goddess Parmeshwara Deva in her manifestation as Brahmari Devi. (Digital image after a modern Hindu style.)
© *Amanda Henriques*

NORTHERN GODS

28 Again, information drawn from Ransome, p46 *passim.*

29 Carved by Teofilis Patiejunas and Ipolitas Užkurnis, *In Your Pocket City Guides.*

30 Andrew Gough 'The Bee' on his *Arcadia* website. Also www.thebeegoddess.com.

31 Hon. J. Abercromby (1889) 'The Beliefs and Religious Ceremonies of the Mordvins', in *Folklore Journal* vol.vii, pp65-135.

Even in northern climates bee gods were worshipped, their blessings in those brief summers being occasions for particular gratitude. Lithuanians venerated the bee-god Bubilas, a fat, hairy figure, and his dainty wife Austeia, a goddess of fecundity, whose statues can be seen in the Lithuanian Museum of Ancient Beekeeping.[29] As a household god he was honoured with honey to encourage swarms to settle in their trees rather than those of their neighbours![30] According to legend, people would move with the queen bee and her swarm to their new hive, binding families together in a special relationship. Anyone finding a dead bee must bury it immediately. Hives were placed in treetops to be nearer to these gods. In common with many cultures, bees and their by-products were considered gifts, not to be bought or sold.

Ancient beliefs survived into the nineteenth century among the Mordvins[31] living west of the Volga. Their Creation myth tells of an invisible, eternal Supreme God,

Cham Pas, who threw presumptuous Shaitan into a sea of eternal fire. Ange Patyai is the life-giving Mother Goddess, whose first son, Nishki Pas, god of sky, sun, fire and light, was also known as the Beehive God around whom good souls cluster like bees round their queen. Her second son protects villages, the 'world-forest-beehive community'; her eldest daughter tends an earthly beehive and protects beekeepers. Another son rules a dark beehive where souls are judged and consigned to Nishki Pas or Shaitan.

There is constant rivalry between Ange Patyai and Shaitan. Her spirits, including a Nizhki *ozais* for each beeyard, protect all aspects of life, but Shaitan sends matching destructive forces. Her identity has become enmeshed with that of Christ's Mother. Producing offspring every day makes the bee Mary's favourite insect, so mead and beeswax candles are essential to her feast days.

Norse stories of Odin's[32] quest for mead are recounted in mediaeval Snorri Sturluson's *Prose Edda and Heimskringla* (see Fig. 3-3). These Viking tales mix myth and history, with sometimes confused origins in a previous *Poetic Edda* and Proto-Germanic beliefs in a Mercury-like god, Wodanaz.[33] Underlying Snorri's generally masculine heroics, this older German belief system featured important goddesses – Erda (Earthmother), Freya, the Norns, Frigga. This world in which Odin had to consult his wives and mistresses, sybils or *haegtessa*, is probably most widely known nowadays through Wagner's *Ring of the Niebelungen* but Tolkien's *The Lord of the Rings* and Rowling's *Harry Potter* series draw on the same sources.

Mead is a divine commodity bringing gifts of wisdom and poetry, which arose when the warring gods Aesir and Vanir both spat in a vessel to conclude a truce. A very wise creature, Kvasir, emerged, only to be murdered by dwarfs and have his blood fermented with honey. Odin's task is to recapture the resulting liquor, Ódrerir, from its guardian, the giantess Gunnlöd. In snake form, he follows a magic tunnel deep into a mountain fastness, seduces her and drinks all this 'mead of poetry'. Escaping in the form of an eagle he spills some to earth where mankind picks it up.

Christian elements crept into the account.[34, 35] Having stolen the mead and left Gunnlöd grieving, Odin hanged himself for nine days on Yggdrasil, the Nordic tree of life, before resurrection.

Several contemporary hives have been very questionably built on these fragmentary Icelandic foundations. Gough's *Arcadia* website draws imaginative connections between bees, Norse ideas and the astonishing but unfinished Rosslyn Chapel in Midlothian. Certainly its founder, William St Clair, was descended from a ninth century Norse earl and its Tree of Life pillar with eight dragons around its base may represent Yggdrasil, a view cautiously endorsed by Wikipedia. A less plausible theory claims significance for a 'curious stone beehive' in the vault nearby but there are many extravagant pendant bosses. Nests recently discovered in other pinnacles suggest no more than bees' natural tendency to squat opportunistically.

The Finnish *Kalevala* clearly blends archetypal fairy tale ideas with religious elements. In one (see Chapter 12), a bee flies far into the cosmos to Ukko himself for a honey salve to resurrect a youth.

As in Greece, these Northern tales link divinity, honey and eloquence. Additionally, bees feeding on Yggdrasil's honeydew connect Odin's identity as a god of resurrection with other traditions where the insects' winter inactivity and spring renewal suggest the migration of the souls of the dead. So many juxtapositions

32 Othin, Wotan

33 According to Roman Tacitus.

34 A section of *The Elder (or Poetic) Edda.*

35 Carolyne Larrington, 'Scandinavia', in Larrington (ed. 1992) *The Feminist Companion to Mythology*, p140.

Fig. 3-3 : Title Page of Snorri's *Edda*
© *Icelandic National Library*

suggest such ideas may have been carried by migrating Celts but perhaps they are essentially common artefacts of a universal human psyche.

CHILAM BALAM AND THE SIGNIFICANCE OF FOUR

Bees had strong religious significance in the distant Americas. Before Europeans brought the honeybee and Latin alphabet, hieroglyphics tell of Mayan people both hunting honey and keeping bees. *The Books of Chilam Balam*, now known as the Madrid Codex (see Fig. 3-4), record their ritual significance; hieroglyphs show gods seated before burning censers and pots of honey.

Living shortly before the Spanish conquest, the prophet Balam told of gods with various responsibilities ranging from rain to death. Four of these, the Bacabs, hold up the four corners of the sky and guard agriculture. Four colours – red, white, black, yellow – associate them with four types of wild bees and the four compass points. The insects purvey messages and food between gods and humans; they feed infant gods; people make offerings to the Bacabs and drink honey wine in their honour. Seeking a blessed abundance for the bees, flowers of the four symbolic colours feature in a festival during the month of Zec.

Not only the seasons but a complex calendar governed beekeepers' activities, restricting them to days called Caban; for instance,

The good day is Thirteen Caban;
The beekeeper is born. [36]

Selima Hill laments this long-forgotten bee-god in a 1984 poem:

> *If ever I find him — thin,*
> *Justly offended, dead*
> *In the dry chaparral —*
> *I will put jade beads*
> *And honey on his tongue…*
> *I will light him candles*
> *Of beeswax, bringing sleep,*
> *And he will rest in the shade*
> *Of the First Tree,*
> *And wait for me there —*
> *Humming a tune, and drinking*
> *Cocoa sweetened with honey.* [37]

A curious detour from this broad religious highway takes us to China, a culture with no conception of a metaphysical universe, but where bees mattered as omens. An early western traveller's brief anecdote adds a Christian gloss. Swarms signalled propitious times to marry, start a building scheme or defer uncompleted business; when he checked a hundred such events against his calendar he was struck by bees' wonderful mystic sense to swarm on Christian holy days! [38]

36 Victoria R. Bricker and Helga-Maria Miram (2002), *An Encounter of Two Worlds: The Book of Chilam Balam of Kaua*, p177.

37 From 'Elegy for the Bee-god', in Selima Hill (2008) *Gloria: Selected Poems* Bloodaxe Books.

38 Retailed by Ransome.

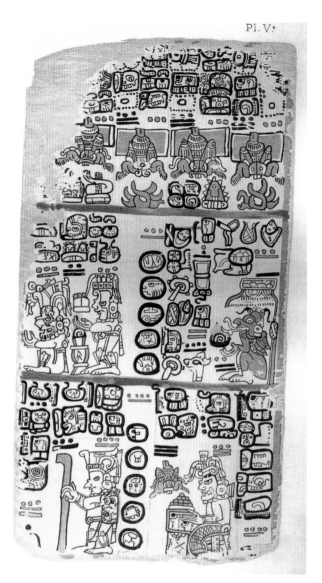

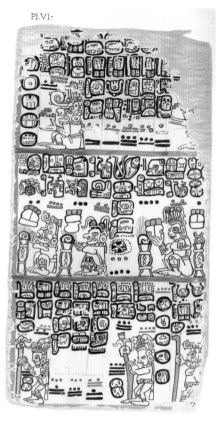

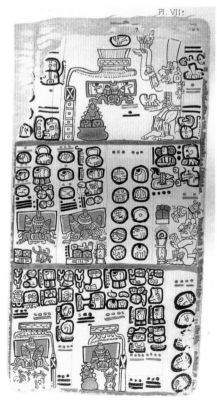

Fig. 3-4 : Celebrating the Honey Harvest, frames from the Madrid Codex

© *Famsi, Foundation for the Advancement of Mesoamerican Studies Inc.*

PART TWO

GREECE AND ROME

Originating in a time beyond recall, Greek myths express our endless enquiry into the nature of existence, telling of a multiplex Olympian realm. Bees may not live there but often feature in rituals, oracles and the gods' earthly manifestations.

Fig. 3-5 : The Nurture of Jupiter, by Nicolas Poussin
© *Dulwich Art Gallery, London*

They link directly to the supreme gods in the Golden Age. Zeus (Roman Jupiter) was born of Rhea in a Cretan cave, hiding from his murderous father Kronos. There he was raised by bee nymphs Melissa and Amalthea, daughters of the island's mythic King Melisseus, and a goat (also known as Amalthea!). Thus he is endowed with the title Zeus Melissaios.[39] When Zeus was grown, these daughters were the first priestesses to the Mother Goddess.

Grateful for having escaped his father's jealous wrath, Zeus gives bees their sting. (Cretan bees are renowned for their savagery.) But then:

39 All these names with the stem 'mel' or 'mal' testify to their bee-laden meaning.

> *Banging about Olympus one sultry day*
> *In another tantrum with his wife,*
> *Perhaps he trod on one unseen*
> *Provoking her sisters to attack.*
> *His outraged majesty decreed*
> *They never will forget his wrath*
> *But die as soon as they unsheath.*

Zeus in his turn fathers a son on a nymph. The boy is abandoned in the woods for fear of his wife Hera's fury. While tending his flocks, Apollo's son Phagros finds the lusty bee-nursed child. Named Meliteus, he later becomes an important king.

LOVE OF SOOTHSAYING

Gathered from wild hives, honey supplied the most primitive of intoxicants, long before corn or vines were farmed or the arts of brewing and winemaking mastered. The motif recurs of singers and seers being fed by bees; Muses sometimes appear in swarms. So long as they can access honey, bee nymphs give Hermes, the gods' messenger, his prophetic gift.

> *For there are sisters born, called Thriae, maiden things,*
> *Three are they and they joy them in glory of swift wings,*
> *Upon their heads is sprinkled fine flour of barley white,*
> *They dwell aloof in dwellings beneath Parnassus' height,*
> *They taught me love of soothsaying, while I my herds did feed,*
> *Being yet a boy. Of me and mine my father took no heed,*
> *And they flitted, now this way, now that, upon the wing,*
> *What time they fed on honey fresh, food of the gods divine,*
> *Then holy madness made their hearts to speak the truth incline,*
> *But if from food of honeycomb they needs must keep aloof*
> *Confused they buzz among themselves and speak no word of sooth.* [40]

Harrison comments that primitive religion aimed to placate gods and ward off evil spirits. Also richly woven into the myths are themes of natural sexuality, manifest in bees' swarming life force, and religious chastity signalled by their inexplicable reproductive habits. But gradually beliefs shifted toward more human images.

GREAT MOTHER, AMBIGUOUS ARTEMIS

By the fifth century BC, the nameless Great Mother Goddess had metamorphosed into the ambiguous figure of Artemis, both virgin and mother. She was the divine huntress who loosed her lethal hounds on the human huntsman, Actaeon, when he stumbled upon her bathing in the woods. Superseding ancient sanctuaries, Cretans built a temple at Ephesus for Greek colonists, where celibates served an important cult in her honour. Resonating with bees and religion, the place links at least the Bronze Age, mythic King Melisseus and mainstream Classical Greek practices.

40 Jane Ellen Harrison (3rd Ed. 1922). The 'Homeric Hymn to Hermes' in her *Prolegomena to the Study of Greek Religion*, pp91 and 442f.

St Paul said the people believed her statue, hung about with amber ornaments, had fallen directly from Jupiter.[41] Interpreted as breasts, eggs or, according to one theory now dismissed, bulls' testicles, one cannot doubt they are potent symbols of fecundity and nourishment and connect readily with the primal Mother idea. Yet there are bees on her flanks and her demand for celibacy clearly links with the enduring enigma of bees' procreation, an association that clung on in western culture until recent centuries.

A second irony is that bees' care of the infant Zeus led to the Mother Goddess's eventual deposition and secured his supremacy in patriarchal Greece's Pantheon.

41 According to a comment in the New Testament, Acts 19.35.

Fig. 3-6 : Roman replica of statue of Artemis, in Museum of Efes, Turkey. Original was destroyed by iconoclastic early Christians
© *Rufus Reade*

Fig. 3-7 : Detail of bees on Artemis' flank
© *Rufus Reade*

42 The title 'sybil' for the several oracular priestesses derives from Cybele, one of the names of the Great Mother Goddess.

43 http://en.wikipedia.org/wiki/Omphalos; Anne Baring & Jules Cashford (1991) *The Myth of the Goddess: Evolution of an Image*; Archaeological Receipts Fund Directorate of Publications, Athens. www.iap.gr; http://www.ancientsites.com/aw/Article/1170885#

44 Lines from the Third Chorus. Translated by E.P. Coleridge (c.1910).

A stone *omphalos* (earth navel) at Delphi marked the sacred place where two eagles met, having been dispatched by Zeus to circumnavigate the earth.

Here Apollo's chief priestess was the sybil Pythia or 'Delphic bee'.[42] Some scholars have suggested the original stone amplified her oracular pronouncements and that its shape and reticulated pattern resemble a woven hive; goddess cultists have claimed a likeness between the surface pattern and bees. While the Delphi museum catalogue makes no such associations,[43] *omphaloi* were also found in other sites with bee connections – the Melissae's original Cretan home and at Paphos, Aphrodite's tomb. She, as a portent of Love's tragedies, has maiden attendants. In Euripides' *Hippolytus* they sing:

> *All things she doth inspire, dread goddess, winging her flight hither and thither like a bee.*[44]

Fig. 3-8 : Beehive shaped omphalos, Delphi, Greece
Courtesy of Mark Cartwright, Ancient History Encyclopedia. www.ancient.eu.com

Virgin priestesses, again called Melissae, serve the harvest goddess, Demeter, at her shrine in Eleusis. Highly secret Mysteries mourn the seasonal abduction of her beloved daughter Persephone into the Underworld and welcome her return in spring. Sacred honeyed foods are presented to the goddess and shared with the worshippers.

In another story of infanticidal threats and marvellous rebirths, Dionysus[45] too was hidden in a cave and raised on honey by a nymph. Again, Zeus is protecting his love child from his jealous wife Hera. His nursemaid is daughter of the pastoral deity Aristaeus, who taught mankind the art of beekeeping. In one manifestation, Dionysus takes the form of a bull that is torn to pieces, only to return to life as a bee!

This indestructible and joyous figure is credited with discovering honey, when 'winged things as yet unknown' are, according to Ovid,[46] drawn by the tinkle and

45 Known by the Romans as Bacchus or Liber.

46 Ovid *Fasti* Book 3.

Fig. 3-9 : Piero di Cosimo 'The Discovery of Honey', c.1505-10
© Art Museum, Worcester, MA, USA

blare of brass and then tracked back to their great honey store. His notorious festival is around midsummer, when the wild honey harvesting is over and Sirius rises in conjunction with the sun.[47] His maenads, bearing honeyed wands, drinking mead and sacrificing a bull, enjoy ecstatic rites in the mountains. His satyrs clash cymbals[48] to attract swarms, whose loud humming seems the voice of the Great-Mother, the sound of creation. Gathering the swarms, Dionysus becomes the father of beekeeping. In further layers of meaning, the swarms are seen not only as the bull's resurrected form but also as rising clouds of undefiled souls.[49]

Ivy-wreathed, Dionysus is surely a primal figure of cyclical nature and a forerunner of our familiar Green Man. Euripides' play *The Bacchae* starts with a benign spring picnic on the mountains until the sexually tormented and puritanical king Pentheus imagines the revels are a lecherous sacrilege. Wise Tiresias advises him to:

> *… welcome this god to Thebes,*
> *Offer libations to him, celebrate his rites,*
> *Put on his garland. Dionysus will not compel*
> *Women to be chaste, since in all matters self-control*
> *Resides in our own natures…*[50]

Rejecting this advice and setting out to smash the celebrations, Pentheus provokes the nature worshippers to a frenzy that ends in his death, seeming like Artemis to express an important transition from earthy sensuousness to the repressed sexuality of succeeding centuries.

FEEDING THE GODS

From Homer onwards we read of people pouring libations on entering temples or before meals, in thanksgiving after battle or during funeral ceremonies. Sophocles' *Oedipus at Colonus* features a very precise rite with holy water and honey to secure the gods' permission for the accursed hero's burial on Athenian soil.[51]

My poem summarises the second century AD geographer Pausanias' accounts of variously exacting local cults.

47 A heavenly conjunction that marks the time for Classical writers to harvest their hives.

48 A folk belief that noise and howling attracted swarms persisted through Virgil until recent centuries. See Chapter 8.

49 An idea that chimes with the belief in bees arising from carcases.

50 Philip Vellacott's translation (1953) lines 314-7.

51 Robert Fagles' translation (1984), lines 519-569.

HONEY FOR THE GODS

People pour libations of mead
And feed their local gods
With parochial honey cakes.
In return they sometimes hear
Promises of lands generously
Awash with milk and honey.
.

Mount Helicon's grass is good
And bees prosper. They praise
Priapus there and hope
For healthy children to laugh and play.

At an arid earthquake crack,
Athenians pour flowing tributes
Into the lost spring, seeking to pacify
Implacable Cronos and Rhea.

Sosipolis demands chastity —
An old woman beyond desire
Must tend his sanctuary
With her dry cakes and clean water.

The Eumenides, Kindly Ones, wait
In an evergreen Corinth grove
For offerings of a ewe in lamb
And garlands with their honey.

Eleusinian bees celebrate
Deeply hidden mysteries;
Each spring Persephone
Returns the fields to work.

Apollo is the most prescriptive god —
To seek his oracle one first must drink
Waters of Lethe and of Memory,
Then dress in ribboned tunics and local boots
And, clutching honey cakes, descend a shaft
Feet first into a river chute.
There might not be a message but
The virtuous survive.

THE ROMAN PANTHEON

The vast Roman pantheon embraces Greek gods under new names. Diodorus Siculus tells how, to preserve an eternal memory, Jupiter (Zeus) coloured his bee familiars like gold-washed brass. He also made them insensible to cold so they could survive the wild weather of their original home on Mount Ida's heights. Diana (Artemis) retains her virginal identity and bee connections, and an obscure minor goddess, Mellona, appears as patron of bees and beekeepers.

4 : CHRISTIAN BEES

THE BIBLICAL HERITAGE

Around 1300 BC Moses led the Hebrews through many hardships from Egyptian captivity toward the Promised Land. The first Commandment brought down from Mount Sinai was 'You shall have no other gods but me', but older beliefs and imagery lingered on. The nineteenth century Egyptologist, Sir W.M. Flinders Petrie, noticed striking similarities between the description of the Jews' sacred Ark of the Covenant and the Egyptians' ark sheltered by the goddess Ma'at. She was the daughter of Ra, whose tears were bees. Trudging across Sinai, Moses' band probably encountered the Ethiopian belief that the bee defended God's throne; some even speculate that their hum could be Yahweh's secret name.[1]

The Bible leaves no doubt about honey's blessed significance in the Promised Land.[2] God's word is often 'sweeter than honey' and two important women bear the name Deborah (bee). Honey and wax candles are important in Jewish ritual. According to Pausanias, the fervent Dead Sea Essenes, thought by many to be forefathers of Christianity, were beekeepers and priestly officials.

Muslims, also 'children of Abraham', revere the bee; true believers are like bees among the most beautiful flowers. The Sunna records that, when Mohammed encountered Jesus in a vision, one of the goblets Jesus gave him contained honey.[3] An-Nahl is instructed in the Qu'ran (16.68-9) to collect honey for mankind's benefit. Also:

> *Rivers of milk, wine and clearest honey*
> *Flow in Paradise. Alone of Allah's insects,*
> *Bees will be found in heaven.*[4]

IMMANUEL. BUTTER AND HONEY SHALL HE EAT...

1 Andrew Gough's *Arcadia* website.

2 Though apparently their earliest 'honey' was a thick grape or date syrup, the Hebrew word being similar to the Arabic *dibs*. *Encyclopaedia Judaica* 2007.

3 Hilda M. Ransome (1937) *The Sacred Bee*, p72.

4 Qu'ran Sura xvi, Sura xlvii.

5 I am indebted to Austin E. Fife (1939) *The Concept of the Sacredness of Bees and Wax in Christian Popular Tradition* for much of this section.

6 Samuel Purchas (1657) *A Theatre of Politicall Flying Insects*, p144. Their Greek names were Persephone and Demeter.

Those words of Isaiah (Isaiah 7.14) are traditionally interpreted as prophesying Christ's birth. Christianity emerged from Jesus's Jewish roots but the New Testament mentions no bees and honey features very little. Like any ascetic preaching in the wilderness, John the Baptist eats locusts and wild honey (Matthew 3.3). Endorsing bees' mystical meanings, Christ chooses to eat a piece of honeycomb to prove his bodily resurrection (Luke 24.39-43). In the Apostle John's visionary book, an angel tells him that God's book would be 'as sweet as honey in thy mouth' (Revelation 10.9).

However, developing amid Graeco-Roman culture, Christian ritual gradually and deliberately added layers of meaning to ancient Israel's milk and honey hopes. Thus, Fife suggests, pagan Greek supernaturalism infused Jewish symbolism and subsequent Christian sacraments.[5] Purchas says the feast of Candlemas derives from the Roman appeal to Proserpine, daughter of the harvest goddess, Ceres, to return from her winter captivity in the Underworld.[6]

Reborn as Christians by baptism, converts were welcomed with a special Mass where five elements – water, milk, honey, wine and bread – were consecrated and shared, symbolising both bodily and spiritual food. Honey and the bee's sting represent both Christ's sweetness and his pains.

That baptismal rite continues in Coptic churches but in the west priests were forbidden 'all carnal sweetness and pleasure'[7] and Saint Thomas Aquinas apparently condemned the usage as primitive and idolatrous. So these antique benedictions became simple thanksgiving for daily food but a festival indulgence at Easter.

DIVINE LIGHT

And so wax overtook honey in western ritual use. Lamps and candles may originally have simply lit Christians' secret meetings in the dark Roman catacombs and some Early Fathers condemned any ritual use as pagan. However, by the time persecution stopped, their symbolism was firmly established. Saint Jerome saw 'no harm if ignorant and simple people or religious women light candles'; such offerings being as acceptable to Christ as Mary Magdalene's ointment (John 12.1-11), and lights express joy during the gospel reading.[8] While Jerome considers a hymn praising candles a pagan excess, Saint Augustine happily composed one! Bees were commonly believed to come from Paradise, and it was fitting that they provided light for the redeeming Mass. Thomas de Cantimpré held that they link human worship to bees' virtue. They acquired specific meanings:

7 Purchas, p143.

8 Charles Herberman (ed) (1913) 'Early Christian Lamps', in the *Catholic Encyclopaedia*.

9 Illuminated mediaeval *Exultet Rolls*, such as a famous one in Bari Cathedral Museum, Italy.

> *The white wax is Christ's Virgin-born flesh,*
> *His soul the wick threaded through*
> *To light the world with divinity's flame.*
> *Phallus-like, the Paschal candle*
> *Dips in the womb waters of baptism*
> *To offer eternal life.*

Nowadays processional candles greet all the great feasts, light the altar during services and flank the coffin in funeral rites. In mediaeval excommunications, priests would terrifyingly invert them and stamp out the light on the ground.

The following formal hymn of praise includes bees themselves as crucial symbols of the Resurrection.[9]

> *Accept this Easter candle,*
> *a flame divided but undimmed,*
> *a pillar of fire that glows to honour God.*
>
> *(For it is fed by the melting wax,*
> *which the mother bee brought forth*
> *to make this precious candle.)*
>
> *Let it mingle with the lights of heaven*
> *and continue bravely burning*
> *to dispel the darkness of this night!*

Fig. 4-1 : Easter midnight vigil: Christ is risen! Passing on the light
Courtesy of Rosie Caldecott and the Oratory, Oxford

Another intriguing link between pagan and Christian religion comes from Russia, where the old bee god Zosim invented agriculture. Along with Saint Sovvatiy, obscure Saint Zosima is the beekeepers' patron, credited with fetching bees from Egypt to remedy the church's desperate shortage of candle wax.[10] However insubstantial the man, he is revered in Russian and Ukrainian Orthodoxy.[11] The bee season starts on his 30th April feast day and Sovvatiy's ends it on 10th October, with prayers to save their bees, gather good honey and not to leave. Ukrainians begin their bee chores by hanging icons among their tools and praying over their hives; the Slavonic liturgical *Book of Needs* provides blessings for all aspects of beekeeping.

And now for a personal contemplation of candles' spiritual significance.

COSMIC CANDLES

The unpolluted dark of the Karoo
Hosts a probing eye prosaically named
The Southern African Large Telescope.
I'm told it would detect a candle on the moon,
Igniting the first of several flames in my mind —

Who would light that candle? An astronaut
With a paraffin wax pack from his kitchen drawer
To insure against power cuts or homesickness
And wing a prayer to an unseen God?
It would take a miracle to flare up there.

So is it God Himself burning
Sacred beeswax in that airless space
Where any trace of atmosphere disappears
For lack of gravity, warming cold hands
And warding off his infinite aloneness?

I'm also told this eye can read across
Ten billion light years — how far is that
In human footsteps, and what messages
Are there at journey's end?
Yet more distances to go…

EXEMPLARY BEES

10 Ransome, p258.

11 Ukrainian-orthodoxy.org/ saints; personal communications.

12 Boundaries are inevitably unclear, especially in the many centuries when Christian belief permeated society at all levels.

As time goes by, bees' virginal, civic and industrious virtues come to the fore, themes evidently valuable to a monastic community but also impressed upon the laity, themes which infuse not only theology but also political and moral philosophies and folklore.[12]

Saint Ambrose (died 397 AD), lawyer, bishop and prefect of Milan, is one of the 'four doctors' of the church. Bees buzzing around his cradle were said to have blessed him, like so many Classical Greeks, with honeyed eloquence. He somehow became the patron of beekeepers, frequently portrayed with a skep[13] or a legend instructing us to be nourished by ambrosia, the food of angels. His name clearly echoes that divine food and he associates Samson's bee riddle with Christ's victorious sweetness – 'O manifest sacrament!'

Fourth century Cappadocian Saint Basil[14] bids us contemplate the genius of honeycombs both as common larder and object lesson in geometrical strength. His contemporary, Saint John Chrysostom, Archbishop of Constantinople and another Early Father nicknamed 'golden mouth', emphasises the hive's community spirit:

The bee is more honored than other animals, not because she labors, but because she labors for others. Indeed, the bee works unceasingly for the common good of the hive, and obeys without question what sometimes appears to be an inequitable hierarchy.[15]

THE 'INCOMPREHENSIBLE GENERATION OF THE SONNE OF GOD'[16]

The supposed chastity of bees was also much revered. Among extensive writing on the subject, Ambrose counsels that nuns' work should be:

… as it were a honeycomb, for virginity is fit to be compared to bees, so laborious is it, so modest, so continent. The bee feeds on dew, it knows no marriage couch, it makes honey. The virgin's dew is the divine word, for the words of God descend like the dew. The virgin's modesty is unstained nature. The virgin's produce is the fruit of the lips, without bitterness, abounding in sweetness. They work in common and their fruit is common.[17]

Fifth century Saint Augustine puzzles at some length about the problem of virtuous procreation:

He was able to produce the first human beings without parents, and was able to form Christ's flesh in the womb of a virgin, and (to address myself even to the unbelievers) he was able to give offspring to the bees without any sexual union.[18]

Devotion to Christ's Virgin Mother grew in mediaeval times, reverberating with echoes of Mother Goddesses and Greek virgin bee priestesses. The *Exultet Rolls* praise the insects' apparent capacity for virgin birth. In ancient Germany, Mary is often imagined as a honey-giving bee, and hives sometimes feature in depictions of the Madonna and Child.[19] Conversely, Fig. 4-2 shows beekeepers discovering them in a hive that concealed a consecrated Host.[20] The early Cistercian and highly revered model of monastic life,[21] Saint Bernard, was a crucial promoter of the Virgin's role as intercessor for the faithful. He was the last to be given the honorific 'doctor mellifluus' title and, often depicted with a symbolic beehive, wrote of honey as life's plentiful, sustaining sap.

Dominican Cantimpré's encyclopaedia also commends these virginal models of the good life, particularly for clergy and monks; convents or monasteries should be quiet in the evening like a hive. However, recalling Pliny's unflattering comparison of

13 The traditional straw beehive.

14 *Sixth Homily on the Days of Creation.*

15 http://thinkexist.com/ quotation.

16 George Gilpin (1579) *The Beehive of the Romish Church,* p350.

17 Quotation from A.E. Fife (1939), p278-9.

18 *Of the Good of Marriage,* para.2.

19 E.g. in *Books of Hours* and in Matthias Grünewald's *Stuppacher Madonna* (1517).

20 See related stories in Chapter 15, the section on Bees at Prayer.

21 Despite failure in preaching against the Cathars and enthusiasm for the disastrous Second Crusade.

E o uilão foy chamar un crerigo ⁊ lli moſtrou ſcā m̄. na colmēa.

E trouuerō a colmēa ⁊ a poſerō ſobelo altar ⁊ diſſerō miſſa.

women and pigs, both said to be aroused by the colour white,[22] he worries about the particular temptations facing his white-robed brothers!

Less strident misogyny but evident distaste for the sexual act blazes through Bartolomaeus Anglicus' assertion that bees':

> … *maidenhood of body without wem is common to them all, and so is birth also. For they are not medlied with service of Venus, nother resolved in lechery, nother bruised with sorrow of birth of children.*[23]

Richard Rolle's[24] essay about bees (and storks so ungainly they cannot fly!) considers a somewhat broader moral model; coming in 'thre kyndis', the first is 'never idyll', the second destroyed because it 'will noghte wyrke' and the third 'kepes clene and bryghte hire wyngeȝ'. Likewise, godly men should be always busy with prayer, thinking, reading or other good, ballasting themselves against 'ȝe wynde of vanyté and pryde'. As bees fight off robbers, so men should resist devils that would turn their 'hony of poure lyfe and of grace' to worms. Besides bees' obedience to the two supreme commandments of charity, Rolle again sounds the familiar chord: they 'hafe othyre vertus, unblendyde with the fylthe of syne and unclene luste'.

It is a relief to turn to pieces less obsessed with virginity. In 1652 John Gage simply likens faithful observance of the liturgical year to bees sucking honey from the blossoms of God's Word. In his book *Silence and Honey Cakes* (2003), Dr Rowan Williams accepts a diversity of vocations and variety in prayer but declares there can be no honey cakes without bees.

In Paradise

Honey streams through images of heaven and departing souls are often depicted as bees in myth and folklore. Struggling in *The Divine Comedy* to describe his vision of the 'unbodied light' of Paradise, Dante visualises the adoring angels:

> … *like a troop of bees,*
> *Amid the vernal sweets alighting now,*
> *Now, clustering, where their fragrant labour glows,*
> *Flew downward to the mighty flower, or rose*
> *From the redundant petals, streaming back*
> *Unto the steadfast dwellings of their joy.*[25]

Church Artefacts

Given so many meanings, it is not surprising to find bee motifs recurring in church buildings, furnishings and vestments. Papal tiaras are said deliberately to resemble a traditional hive. Ceremonial objects have an obvious symbolic purpose but other depictions may have no more spiritual intent than everyday scenes in Books of Hours.

On the capitals (at the top of columns) in such great churches as Vézelay and Cluny, beekeepers inspect hives; a corbel in St James' Church, Taunton, shows a

22 Mary Carpenter Erler and Maryanne Kowaleski (eds. 2003) *Gendering the Master Narrative*, p42 .

23 Robert Steele (1240; 1905 ed.) *Mediaeval Lore from Bartolomaeus Anglicus*, p124.

24 Fourteenth century Yorkshire mystic, hermit and poet.

25 From Canto XXXI, translated by Henry Francis Cary.

swarm running into a skep; while bees also decorate the Parma Baptistry. Tombs repeat ancient messages of dormancy and resurrection.

Political dimensions become unambiguous in Vatican imagery. The Barberinis[26] were originally peasants named Tafani or 'Horsefly' and these tiresome insects featured on their family arms. However, with the family's social ascent, they were exalted to bees and a palace painting depicts them surrounding the Virtues. When a swarm invaded a seventeenth century papal election it was taken for a sign that a Barberini should be chosen. Then, as Pope Urban VIII, he commissioned Bellini's grandiose baldachin in St Peter's. Bees adorn its sculpted greenery and the family's three armorial bees decorate the plinths and other furnishings. He received Stelluti's presentation plate depicting bees in microscopic detail.

Fig. 4-3 : Papal Coat of Arms of Urban VIII
Courtesy of www.araldicavaticana

Fig. 4-4 : Francesco Stelluti's Melissographia (1625), presentation plate to Pope Urban VIII, an early microscope-aided drawing of bees NLS 744I2466.I
© *National Library of Scotland*

Reformation, Counter-Reformation and Revolution

In all the surge and bloodshed of European Reformation and Counter-Reformation and English Civil War, even beehives deserve an X-rating! George Gilpin was a scion of Westmorland gentry divided between adherents of the old religion and the new and also between Royalists and Roundheads. Converted to Protestantism around the 1550s, he became Queen Elizabeth's agent in Antwerp, a situation requiring delicate diplomacy to maintain England's trading interests and support the Low Countries' newfound Protestant resistance to Spanish overlords while an armada lay offshore.

26 Various websites on the family.

Fig. 4-5 : Beehive of the Roman Church, from title page of van Marnix' 1569 Edition. Shelfmark J-J Sidney 163, Plate Facing Page 350

© The Bodleian Libraries, The University of Oxford

But this fluent Dutch speaker showed no finesse in translating a bitter polemic against the 'Romish Church',[27] authored by Calvinist theologian Phillips van Marnix van St Aldegonde. Presenting himself as a superstitious papist he:

> *... doth so driely resell the grose opinions of popery, and so divinely defend the articles of Christianitie, that (the sacred scriptures excepted) there is not a booke to be founde either more necessarie for they profite, or sweeter for they comforte...*

One begins to wonder whether either the original Calvinist or his translator is a double or triple agent! Purporting to be a good Franciscan and addressing the work to the Bishop of Leeuwen, he wants all good Catholics to understand it. In an English edition of 1579, a sarcastic introductory letter to the Bishop of S'Hertogenbosch thanks him for torturing Huguenots and reducing Truth to 'a muddy bolus [of] sophistical seacoales'. And yet this hive is offered as a model of the Church's former chaste and sober virtue, toughly woven from osiers of Loven and Paris by Duns Scotus, Aquinas and suchlike experts, and bound together by Jewish cables. A refining daub of ancient Doctors and the Council of Trent's new chalk, plus superstitious and heretical sand and cleansing bitumen, seals a weatherproof structure, ready for 'all manner of gallant pictures and images' to attract the bees.

In stark contrast, sectarian conflict distressed Joseph Hall, ordained in 1600 and subsequently Bishop of Norwich. His meditations range from a dormouse to an arm with pins and needles! Singing the usual praises of the hive commonwealth, he relishes the punishment of wasps and drones, 'unprofitable hangbys', but he hates warfare at the hive mouth.[28] This sad and hateful spectacle is like good Christians 'shedding each other's blood upon quarrels of religion', and he prays for peace among brethren.

27 *The Beehive of the Catholik Church* (1571).

28 Joseph Hall (1630) *Occasional meditations*, p150.

Puritan scholars also found inspiration in the hive. In *The Spirituall Use of an Orchard* (1653), Ralph Austen writes:

> *Our eyes ought as little bees fall upon several objects and… gather hony, and bring it into the hive; that is sweet, heavenly, wholesome Meditations for magnifying the Creator in all his Attributes.*

The clerical attire in Figure 4-6 shows the man is a Protestant. He is eating honeycomb and praising pious scholarship.

These exemplary insects continued in lockstep with both Reformed and Catholic pietists. The Counter-Reformation produced such zealous books as *The Christian Sodality* (1652), which offered to help the Catholick hive to pray with informed devotion. In the obsequious manner that seems to characterise the era,[29] the compiler 'FP' presents himself as 'the Puny Bee of all the Hive', undeserving of his full name. Humbly seeking aristocratic patronage, he confesses his work is piously stolen from the Church as openly as bees steal from gardens. But empty recitations, even of 'these sweet honey Combs of Grace', are no substitute for individual sanctity.

During the Cromwellian Commonwealth, Samuel Purchas published his *Theatre of Politicall Flying Insects* (1657). Everyone, even the poor, can have a prayerful spirit within Christ's hive and, like obedient bees, pursue 'a higher aspect'. Surfeiting on either honey or pleasure is wrong, and men should not be drones but forage earnestly for self-improvement. A man without Christ is as helpless as a single bee; God is the strong house; Heaven is 'a full felicitie to the soul'; honey is all honey.

29 A style seen repeatedly in dedications in Chapter 6.

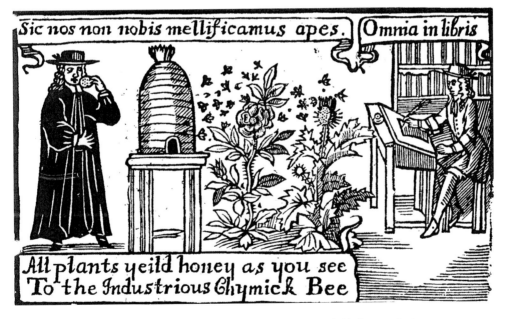

Fig. 4-6 : A woodcut of the late 1600s, showing both roses and thistles producing sweetness. The Latin inscriptions read 'So we the bees make honey, but not for ourselves' and 'Everything is to be found in books'
Found in Eva Crane (1999) World History of Beekeeping and Honey Hunting, p607, no source given

5 : MORAL AND SAGACIOUS BEES

GUIDES TO INDIVIDUAL ETHICS

Both Rolle and Purchas in the last chapter straddle the boundaries between political, civic and social morality and industry. This chapter focuses on bees' place in individual secular morality.

Fulsomely praising bees' inner excellence, Charles Butler catalogued their 'curious art and admirable virtu'. Valiant and magnanimous in battle, they would rather use fangs than their suicidal spears to chide. Such 'great hearts… in little bodies'[1] tolerate personal slights. They forage with wit and dexterity and read the weather skilfully, employing both wind and lee hedges. 'Swifter than the east wind', they labour incessantly until worn out. Then they refuse food in the interests of the hive!

Purchas further praises the insects' industrious, frugal, clean and chaste habits. Concerning their leonine courage, he recalls that the Punic War General Xanthippus would prefer to serve under a bee commander than himself command worthless 'pismires'.[2]

MAKING A BEELINE

The commonplace phrase 'to make a beeline' derives, according to Gough's *Arcadia* website, from ancient Phoenician sailors' practice of releasing bees to lead them to a landfall. He suggests that ships and bees on port city arms may reflect that tradition. But the phrase could equally have arisen from timeless observation of the laden bee's flight home. The earliest printed trace was in an 1808 US newspaper's rather surreal notion of a panicking horse 'making a beeline' for home.[3]

FORAGING FOR WISDOM

According to Plato, Socrates threatens to 'go off like a bee leaving its sting behind' if students missed flaws in his argument. In his view, a virtuous life mixes honey with water rather than seeking unalloyed pleasure, but he acknowledges anger can also be sweeter than honey. Achilles' pleasure in his legendary wrath until the death of Patroclus breaks his heart provides the motivation for Homer's tragic *Iliad*.

Isocrates passionately deplores the cheap rhetoricians debasing Athenian democracy and advises his eager student, Demonicus, to model himself on his father and Heracles but also to attend to poets and wise men:

> *For just as we see the bee settling on all the flowers, and sipping the best from each, so those who aspire to culture ought not to leave anything untasted but should gather knowledge from every source.*[4]

1 *The Feminine Monarchie* (1609) Ch 1, para 48.

2 Samuel Purchas (1657) *A Theatre of Politicall Flying Insects*, p18.

3 Gary Martin, *The Phrase Finder* on www.phrases.org.uk.

4 *Isocrates to Demonicus 52.*

The Roman statesman Seneca also mentors a young friend, Lucilium, commending Nature's reliable supply of survival skills. Bees co-operate and display:

> *… adaptability and self-love. They could not survive except by desiring to do so… In no animal can you observe any low self-esteem, or even any carelessness, of self. Dumb beasts, sluggish in other respects, are clever at living.* [5]

Bees' self-sacrificing and conscientious nature wins Virgil's praise. With torn wings and exhausted:

> *Such rage of honey in their bosom beats;*
> *And such a zeal they have for flowery sweets.*
> *…*
> *The fortune of the family remains;*
> *And grandsires' grandsires the long list contains.* [6]

The familiar Biblical 'Go to the ant, thou sluggard',[7] sometimes carries a Greek gloss, 'Or go to the bee, and learn what a work a woman is…'. The Talmud extols bees' inexhaustible industry and social genius and the Qu'ran singles them out as God-given tutors to mankind. A fourteenth century Arabic Zoological Dictionary[8] asserts that, of all flies, only 'the honey-fly' escapes hell. Another ancient work, *The Wonders of Creation*, added intelligent foresight to the bee's already familiar virtues.

Founding Oxford's Corpus Christi in 1517, Bishop Richard Fox of Winchester thought so highly of the insects' manifold virtues that they provided an elaborate metaphor for the Statutes. The tutors would be skilful herbalists offering a curriculum of the choicest honey-bearing flowers, and the whole College would be:

> *… a bee garden… wherein scholars like ingenious bees, are by day and night to make wax to the honour of God, and honey, dropping sweetness, to the profit of themselves and all Christians.* [9]

It was widely believed that bees would not sting a chaste girl! The ancient and dubious belief that they dislike strong smells became moralised into a precept of sexual abstinence. A sixteenth century Dutch 'prodigy of learning', Joseph Justus Scaliger, declared:

> *Bees sense when a man has slept with his wife, and undoubtedly the chaste creatures will sting him if he approaches them the next day.*

The Puritan Ralph Austen's bee-filled orchard provides 'a very garden of Delight… a Canaan'[10] for illiterate and lettered folk alike, evoking sweet, heavenly, wholesome reflections alongside economic benefits. His manual is therefore a 'plaine, sound, Experimental work' interwoven with Classical, Baconian and religious ideas, and his readers should be neither Drone nor Dunce, setting to real work rather than frittering time away in university!

In 1704, Jonathan Swift[11] could write that bees furnish mankind 'with two of the noblest things… sweetness and light', gifts which Matthew Arnold later marked out

5 Seneca *Epistulae Morales*, Epistle.cxxI,22, 23.

6 Dryden's *Virgil*, p114.

7 Book of Proverbs, 6.6.

8 Kemal-Aldin Muhammad Aldamari *Hayat Alhaywan, or Life of Animals*, quoted by W.C. Cotton (1842) *My Bee Book*.

9 Quoted by permission of the President and Fellows of Corpus Christi College, Oxford.

10 Ralph Austen, (1653) *The Spirituall Use of an Orchard*, Preface.

11 Jonathan Swift (1704) *The Battle of the Books*.

as culture's best attributes. Cotton's 'Letter to Cottagers' proposed a different and peculiarly Victorian moral instruction to the labouring classes. An observation hive provides:

> *Lessons for those who have Eyes! ... PATIENCE, PERSEVERANCE, INDUSTRY, TEACHABLENESS, LOVE TO THE QUEEN; LOVE FOR THE LITTLE ONES; DUTY TO PARENTS, and many others which you may hunt out for yourselves.* [12]

During his retreat to Walden, the American philosopher Henry Thoreau[13] made it his everyday business, by steeping himself in nature, to learn 'how to live… how to extract its honey from the flower of this world'. Alert to flowers' timetables, bees could teach botanists to work faster. More metaphorically, Nietzsche sees human inquisitiveness as a mirror of their quests.

> *Our treasure lies in the beehive of our knowledge. We are perpetually on the way thither, being by nature winged insects and honey-gatherers of the mind.* [14]

TWENTIETH CENTURY AFTERTHOUGHTS

To the Lebanese philosopher poet, Kahlil Gibran, gardens offered living lessons in the difference between good and bad pleasures.

> *... it is the pleasure of the bee to*
> *gather honey of the flower. But it is also the pleasure of the flower to yield its honey*
> *to the bee.*
> *For to the bee a flower is the fountain of life.*
> *And to the flower a bee is the messenger of love.*
> *And to both, bee and flower, the giving and the receiving of pleasure is a need and*
> *an ecstasy.* [15]

In the very same year Rudolf Steiner, the Swiss founder of the benign but often obscure Anthroposophy, strikes a strange note. Although bees' loving commonwealth feeds from flowers' sexual parts, their renunciation of sexuality is a lesson to the human soul. Thus, as their sustenance comes from the plant's love life and reaches us through the tongue's sensual pleasure, they provide the route by which:

> *...the entire cosmos can find its way into human beings and help to make them sound in mind and body.* [16]

OCCASIONAL DISCORDS IN THE SYMPHONY OF PRAISE

Some observers refrain from moral judgements but bees' robber instincts disturb other admirers. Bishop Hall deplores the owner's losses as well as the insects' ruthlessness in falling foul of their own kind, whilst Root[17] bluntly condemns their

12 William C. Cotton (1842) *My Bee Book*, p311.

13 David Spooner (2002) *Thoreau's Vision of Insects*, pp50, 56.

14 http://www.1-famous-quotes.com/quote/72728

15 *The Prophet* (1923; Penguin ed. 1992), p94.

16 Rudolf Steiner (transl. Thomas Braatz) *Bees*, pp3-4.

17 A.I. Root (1908) *The ABC and XYZ of Bee Culture*.

ransacking of a weaker hive. Thorley remarks on their savagery when exasperated around their hives:

> *… you may as well take a bear by the tooth or offer to bind a lion by a single thread or hair as by resistance and opposition to compose and quiet them.* [18]

But Maeterlinck[19] notes their 'strange duality' – masses avenging a single sister injured in the hive, yet showing total indifference to handfuls mutilated outside on a piece of comb. Indeed, the bees will be busy scavenging the leaking honey. Caring only for the future of the hive, in modern parlance this is the ecologically efficient behaviour pattern of a superorganism.

The Drone as Antihero

Ever a byword for worthlessness, drones are calumniously identified by Hesiod[20] as indulged ladies. However, he also chastises a loutish man, Perses, as a drone, a great fool universally resented. Socrates likens them to lazy loudmouths in the Athenian democracy and Aristotle observes they are punished 'as is apt to those who have no function'. The Roman Aelian says drones purloin the honey while weary workers sleep. Xenophon[21] merely calls them 'virulent weeds, to be eradicated'; Christian legends say they are the Devil's creation.

Butler[22] describes a fellow only flying noisily at midday 'to get him a stomach' before returning to his chair and the full combs. The Puritan seventeenth century heaps further opprobrium: 'unprofitable and harmfull Hangbyes',[23] gorging until 'so soggy' they can only crawl from the hive and are unable to fly back;[24] 'idle, sluggish, lither and ravenous cloystered Monks'.[25] Moffett speculates upon their possible domestic and reproductive functions, but concludes:

> *Both God and man disdain that man*
> *Which Drone like in the blue,*
> *Nor good, nor ill, endeavour can*
> *Upon himself to live,*
> *But idle is, and without sting,*
> *And grieves the labouring Bee,*
> *Devouring that which he home brings,*
> *Not yielding help or fee.* [26]

Wildman, too, notes their slugabed nature, only out on fine afternoons collecting nothing. That they return full of honey is, he believes, simply because they took it from the hive in the first place.[27]

Whereas in bad weather a worker would die rather than befoul the hive, says Maeterlinck, drones imprudently soil the combs. In early summer they behave like Penelope's loutish suitors, feasting, carousing and jostling, sleeping in the snuggest corners, knocking workers aside from their jobs to swagger out at midday for a nap in a nearby flower, then returning to the honey vats when the air cools.

However, he cannot but admire their appearance, those:

18 John Thorley (1744) *Melisselogia*, p16.

19 Maurice Maeterlinck (1901) *The Life of the Bee*, pp126-7.

20 *Works and Days*, line 305.

21 *Economics* 17.14.

22 Butler (1609), Ch 4,1.

23 Joseph Hall (1630) *Occasional meditations*, p149.

24 Purchas, p40.

25 E. Topsell, (1658) *The History of Serpents*, p638.

26 Thomas Moffett (1658) *Insectorum*, p892.

27 Thomas Wildman (1768) *Treatise on the Management of Bees*, p31.

... two enormous black pearls, two lofty quivering plumes, a doublet of iridescent, yellowish velvet, an heroic tuft, and a fourfold mantle, translucent and rigid.[28]

Edwardes, too, shows a sneaking admiration for the horseplaying sluggard, stylish with 'his round velvet cap, his side gowne, his full paunch, and his loude voice'.[29] But workers receive his returning fanforanade sullenly, indifferently, and such parasitical behaviour cannot be excused. Despite knowing that their physique explains their great indolence, Von Frisch nevertheless castigates them for waiting for workers to feed them.

In purely metaphorical settings, we find Cantimpré[30] calling lay-brothers drones, which seems unduly hard on those who did all the donkeywork of many a monastic community. In words that should still resonate, Shelley's 'Song to the Men of England' contrasts aristocratic drones with the actual wealth-makers:

> *Wherefore, Bees of England, forge*
> *Many a weapon, chain, and scourge,*
> *That these stingless drones may spoil*
> *The forced produce of your toil?*

28 M. Maeterlinck, p286.

29 E. Edwardes (1908) *The Lore of the Honeybee*, pp43, 242, 243.

30 Thomas de Cantimpré *The Mediaeval Bestiary.*

Fig. 5-1 : Drones, a character study as depicted in Cotton's *Buzz-a-Buzz*. Shelfmark 280 I 136, p7
© The Bodleian Libraries, The University of Oxford

Dickens's charming sponger Skimpole[31] rejected the bee's priggish diligence and preferred the drone's hedonism – 'so much to see, and so short a time to see it in'. Bertie Wooster[32] notoriously tottered from his bed to the day's pleasures at the Drones Club.

A Cubbyhole in my Cabinet

There are said to be more than 120 pub and inn signs featuring bees or hives. Unlike familiar obeisances to King's Heads, King's Arms or the local lord, these have none of the moral, religious or political overtones evident in other bee iconography. They presumably are rooted in the commonplace, reflecting the customers' interests like Bricklayers' or Farmers' Arms. A more extraordinary case is Grantham's Beehive pub, with a colony maintained continuously since a swarm settled there in 1798; this phenomenon is marked by the hive itself and by this plaque on the wall:

> *Stop Traveller! This wondrous sign explore*
> *And say, when thou hast viewed it o'er and o'er,*
> *Grantham, now two rarities are thine.*
> *A lofty steeple and a living sign.*

My devout grandmother and worldlier grandfather lived just round the corner from these two wonders and introduced me to both during childhood holidays.

31 *Bleak House*, p106-7. Oxford World Classics Edition (2008).

32 P.G. Wodehouse (1991) *The Drones Omnibus.*

6 : FOR TYRANNY OR COMMONWEALTH?

ASCENDANCY OF POLITICAL METAPHORS

Different societies at different times interwove strands of sacral power, sexual virtue, sober industry and institutional authority into very diverse patterns to suit their own interests, but from the sixteenth century onwards political metaphors became more prominent. Secular writers, albeit often with a religious gloss, fervently enlisted bees to legitimise or occasionally critique the powers that be:

> Oh wonderful! Hath the All–wise Creator placed such Wisdom, such curious Art, such Fortitude and Foresight, so polite a Government, and such indefatigable Industry in creatures so small. [1]

They appear on ancient coinage and continually on official emblems yet their organisation is governed entirely by instinct, as Aristotle observed, but this did not deter him or many successors from offering the hive as a model for a good society.

GENDER BATTLES

We have seen how much the notion of female authority caused discomfort and complicates the idealisation of the hive as a social model. We cannot miss the misogyny of Hesiod's confusion of parasitical drones with 'the deadly race and tribe of women… no companions of poverty but only of luxury',[2] or Juvenal, for instance, classifying women along with livestock, among whom only the bee-like, asexual, industrious woman is praiseworthy. Proposing a model of household management to his young wife, Xenophon declares such natural roles, endorsed by law, are:

> …of no small moment… unless, indeed, the tasks over which the queen bee in the hive presides are of small moment… God from the first adapted the woman's nature, I think, to the indoor and man's to the outdoor tasks and cares. [3]

On the other hand, Athens' comedic playwright Aristophanes[4] offers ruses for women to evade men's tightening hold on their lives. One trick, when meeting a lover, involves a phantom baby in a honey jar. He even forces the tragedian Euripides to repudiate his poor opinion of women.

Saint Ambrose simultaneously and contradictorily exalts bee virginity and 'the blessed and marvellous mother bee'. Wilson[5] suggests this, like the Anglo-Saxon word *beo-modor*, was no more gender-laden than the phrase Mother Nature but we cannot so lightly overlook the recurring denigration of women's sexuality or entitlement to authority. Many centuries passed before Butler's *The Feminine Monarchie* unequivocally proclaims the truth but even he deems Aristotle's Rex a worthier title and uncompromisingly declares:

1 Joseph Warder (1712)
True Amazons, p162.

2 *Theogony*, line 585ff.

3 *Economics VII*, 17.

4 *The Poet and the Women*.

5 Bee Wilson (2004)
The Hive, p89.

Let no nimble-tongued sophisters gather a false conclusion... [the hive does not license women to] arrogate to themselves the like superioritie. [6]

Women should copy worker bees' virtues and, if they must, only discreetly punish a drone or dull lubber.

Within a few years, microscopist Swammerdam[7] was greatly surprised to confirm this female monarchy but even after Bazin's translation[8] reached Britain in 1744 many still found the idea politically indigestible. Arthur Murphy is an exception; entertaining young Miss Susanna Arabella Thrale to truths unknown even to Dryden with his translation of Jacques Vanière's *The Bees*[9] – 'No Salique-law[10] excludes the female line'. More typically, Jean-Baptiste Simon[11] insists the hive's royal lineage requires a male ruler and female mother. Since male authority was decreed in Eden, John Thorley[12] also deplores the queen bee's dignity and the drones' contemptible status:

... triumphed over and trampled upon by the Populace... and slain without Mercy? Or have these Gentlemen forgot what they were taught when Schoolboys, that the male gender is more worthy than the female?[13]

Neither, reminding us that Eve drove Adam to a 'very foul thing', can John Keys[14] tolerate a dominant female. Unlike many admirers of the queen bee's beauty, he calls her ungainly, a 'tall woman in a short cloak', quite simply (and accurately!) an undignified breeding machine.

Dismay continues into the twentieth century. Could the hive foreshadow a human future, 'an eminently disquieting state of... matriarchy triumphant' with a mere handful of males fulfilling one indispensable office?[15]

UNA APIS, NULLA APIS

Lycurgus, founder of Sparta, is said to have modelled the proverbially martial city on the hive, warning against noisy drones and any who only work for reward. A wise beekeeper must cut out sick cells and prevent a saviour dictator or oligarchies emerging.

'The bee is nothing alone' says the classical proverb. Aristotle is the first of many to admire their domestic hygiene, conscientious foraging until worn ragged, and uncompromising defence of their community. In his bee garden's drowsy shade, Virgil turns such observations into high poetry, a staple of many English poets. In Dryden's version of *The Georgics*, they represent a busy shop, full of trading citizens.

Of all the race of animals, alone
The bees have common cities of their own,
And common sons; beneath one law they live,
And with one common stock their traffic drive.
Each has a certain home, a several stall:
All is the state's; the state provides for all.
Mindful of coming cold, they share the pain,
And hoard, for winter's use, the summer's gain.

6 Charles Butler (1609) *The Feminine Monarchie,* Sections 4.6, 4.23.

7 Jan Swammerdam (1758) *The Book of Nature,* p121.

8 Gilles Bazin (1744) *The Natural History of Bees.*

9 Arthur Murphy (1799) *The Bees: a Poem from the Fourteenth Book of Vanière's Praedium Rusticum.*

10 Murphy, p19. The Common Law of France, which forbade female succession to the monarchy.

11 *Le Gouvernement Admirable* (1740), p.xix 4-14. Workers and drones reproduce themselves!

12 *Melisselogia, or The Female Monarchy* (1744), p90.

13 Thorley, p90.

14 (1780) *The Practical Bee Master,* p2.

15 Edward Edwardes (1908) *The Lore of the Honeybee,* p. xvii.

He lists the workers' various tasks, from building comb to driving out the drones. Another translator likens them to Cyclopean blacksmiths, goaded on by the love of getting: 'How glows the work!' Bees sit on their brood like hens on their eggs, said Cicero.

Virgil was not wholly inaccurate in observing their part in appointing and dismissing their monarch. Differing as to whether the monarch has a sting, all writers agree about his subjects' surprising obedience and protectiveness, surrounding him like guards and lictors.[16]

Mediaeval bestiaries praise this subordination to an elected, revered and industrious king, 'most worthy and noble in highness and fairness, and most clear in mildness'; he receives knightly fealty, is freely loved and mightily defended. It is an honour to perish for him, says Bartolomaeus Anglicus.[17] His contemporary, Cantimpré,[18] sees the hive as a model for mild bishops' rule of the church.

More complex beliefs about hive ranks (see Chapter Seven) did not long survive scientific study but many of those notions endured as models for human society. The Latin tag *sic vos non vobis mellificatis apes* (so do you, bees, make honey, but not for yourselves) became a sentiment of *noblesse oblige*. Sir Philip Sidney and other benevolent aristocrats adopted a beehive emblem and the motto *Non nobis*. But the hive has served to endorse constitutions ranging from absolute but benign monarchy to an ideal commonwealth.

16 A lictor was the ancient Roman official who carried a fascis (bundle of rods) and attended magistrates.

17 Bartolomaeus Anglicus *Of the Properties of Things.*

18 Thomas de Cantimpré *The Mediaeval Bestiary.*

Fig. 6-1 : Detail of obelisk showing bee and reed symbol, now in Istanbul, originally from Thebes, Egypt, c.1500 BC
Courtesy of Rufus Reade

On the Side of Human Monarchy

Bees being the god Ra's fruitful tears, the Pharaohs adopted one as their symbol of kingship. It mingled the sting of justice with honey's clemency, and provided a model of hierarchy and productive work. Around 5000 years ago, when King Menes united the Upper and Lower Nile kingdoms, the Kahun Papyri record it as the joining of the Reed to the Bee. Men-kau-ra's coffin fragments in the British Museum bear the symbol, and it recurs in pyramids, on obelisks and tombs. Later, in the XVIIIth Dynasty, bee ornaments were given as medals for valour.

Ancient tradition had it that bees guided the Ionians to settle in Greece, whilst swarms provided Homer's simile for Agamemnon's armies mustering to fight Troy.

Was Seneca diplomatic or sardonic in advising his student Nero to contemplate the hive's monarch as a model? Quite properly, the king bee is brightly coloured and lives in luxury, merely supervising others' work, but his worthiest distinction is to have:

> *...no sting. Nature did not wish him to be cruel or to seek a revenge that would be so costly, and so she removed his weapon, and left his anger unarmed.*[19]

Whatever the respected statesman's frame of mind when writing this, it was a cruel irony that Nero later ordered him to commit suicide.

Affirming monarchy as the natural order, Saint Basil not only dismisses election by an ignorant populace but also blind fate and heredity, all being likely to enthrone the most unworthy. So he considers by what mechanism 'nature' in the hive does better.

> *It... gives him superior size, beauty, and sweetness of character. He has a sting like the others, but he does not use it to revenge himself. It is a principle of natural and unwritten law, that those who are raised to high office, ought to be lenient in punishing.*[20]

19 *Of Mercy,* ch.19.1-3.

20 *Sixth Homily on the Days of Creation.*

However, wrathful, imprudent bees will promptly repent, dying by their own stings!

Claiming a blood line back to the Holy Family, early French kings adopted the bee as a symbol of holiness and royalty. Childeric I, the fifth century Frankish warrior reputedly wore a coat decorated with bees and bearing the motto 'The King wants or useth not his sting'. When repairers opened his tomb in a Tournai church in 1653, over 300 gold bees were found, among other treasures. Their later chequered history deserves retelling. They passed through the Habsburg dynasty in Vienna before returning to France, whereupon Louis XIV stowed

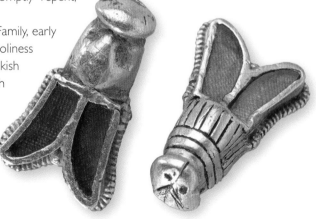

Fig. 6-2 : The two surviving bees from Childeric's tomb INV-55-449-50

© *Bibliothèque Nationale Française*

them in a library drawer. The 1789 Revolution brought them to the Bibliothèque Nationale until, seeking to outrank the Bourbon fleur-de-lys, Emperor Napoleon had them sewn on his coronation robes! But further misadventures befell. After his downfall, the canons of Nôtre Dame sold the bees by the pound to repair their roof. Then, among much other treasure stolen in 1831 and melted down or hidden in the Seine, only two survived to return to the Bibliothèque.

It has been suggested that the fleur-de-lys was itself a disguised (or perhaps merely badly drawn) bee, adopted by the court during eleventh century persecution of the Cathars. However, twelfth century royal propaganda declared that the Blessed Virgin presented the lily at Clovis I's baptism. Whatever the truth about the fleur-de-lys, bees were enduring regal symbols. With Childeric's motto they were used to decorate Louis XII's clothes when he fought the Genoese in 1506,[21] and were later appropriated by Napoleon.

Butler fulsomely dedicated his book[22] to Her Majesty, citing the queen bee as 'the most ancient and invincible Monarch' governing an exemplary society that equally abhors polyarchy and anarchy. Followed by the customary praise of bees' communal virtues, he hymns her:

> *… entire and absolute power, unsurpassed beauty, majesty, temperance, taciturnity, and other princely feminine graces…*

loved, revered and obeyed in all things by her subjects.

> *If she goes forth to solace herself many attend her;*
> *If her voice bids them swarm, they obey;*
> *if then she dislikes the weather*
> *or lighting place they quickly return;*
> *if she directs, they fight.*
> *While she is well, they enjoy their work;*
> *if she droop and die, they too*
> *will languish and die.*

So much for the kingly ideal – justice and leadership needing no sting – and dutiful subjects. Complexities arise when the king fails to meet expectations.

While Erasmus was commending the king bee to the Spanish *Rey Cattolicos* Charles V, Sir Walter Raleigh was comparing the newly Protestant and capricious Henry VIII to a bad beekeeper:

> *To how many others of more desert gave he abundant flowers from whence to gather honey, and in the end of harvest burnt them all in the hive!*[23]

Shakespeare berates young Harry for courting popularity among 'cap'ring fools' sickening themselves with surfeits of honey. The lad repents and emerges as the inspiring Henry V, ruling over an orderly working hive with its various officers, soldiers, mechanics and porters.[24]

John Day's satirical *Parliament of Bees*[25] depicts a hive containing twelve combs. His characters range from Thraso and Pharmacopolis, who respectively 'grind the faces of the poor' and 'steal practice and buy patients', to an upright soldier and an

21 Jeffrey Merrick (2007)
Order and Disorder under the Ancien Régime, p3.

22 *The Feminine Monarchie* (1609). The date of publication and this dedication present a belated endorsement of Elizabeth I's reign; she died in 1603 and was immediately succeeded by James I.

23 From the Preface to Raleigh's *The History of the World*.

24 *Henry IV Part I*, Act 3, 2.1; *Henry V*, Act 1.2, Archbishop of Canterbury's closing speech.

25 John Day (1641) *The Parliament of Bees*.

honest broker. Seeming to mock the fashion for over-egged flattery of royalty, he dedicates it thus:

To the worthy Gentleman Mr George Butler, professor of the Arts Liberall, and true Patron to Neglected Poesie, All Health and Happinesse.

Worthy Sir,
I may be thought bold, if not impudent, (upon so little acquaintance) to make this sawcy trespasse upon your patience; But fame, whose office, (like the Nomenclators at Rome) is to take notice and proclaime the Name and Vertues of every noble Personage, has given you out for so Ingenuous a professor of the Arts, and so bountiful a patron of poor schollars it has imboldned me, to present my hive of Bees to your favourable protection; and when I remember how Lewis the eleventh (of that Name) King of France tooke notice, & bountifully rewarded a decay'd Gardine, who presented him with a bunch of Carrets, I doubt not of their kinde and generous entertainment; upon which assurance I rest ever.

Yours in all service devote.
John Day.

Thomas Mouffett[26] unfavourably contrasts courtiers and kings who pursue their own lusts with 'these little winged beasts' who only offer allegiance to a sovereign circumspectly chosen.

REFORMATION, COMMONWEALTH, RESTORATION

To retain your head during the sixteenth and seventeenth centuries required diplomatic agility. Prudence dictated whether bees embodied monarchism or commonwealth while power was tossed between monarchs favouring the new Reformed Church or allegiance to Rome, and then between kings and commonwealth. The tussles continued until finally William of Orange's Protestant and constitutional monarchy was established in 1688.

While one party claimed the hive model of kingship as an antidote to any individual's actual frailties, Cromwell asserted it symbolised parliamentary rule. Published during the Commonwealth, the main message of Samuel Purchas's *A Theatre of Politicall Flying Insects* is about earthly governance. Citing Marcus Aurelius's 'What is not good for the swarm is not good for the bee', the hive's commonwealth proves that 'a Christian King should abase himself'. Whose side is he on? Elsewhere he commends its queenly realm and cautions mobs against swarming after ambitious leaders; the hive's laws are natural, 'graven in their manners; and so studious… of peace.'[27]

Some bees themselves voted with their feet during the Civil War period! When a Parliament Visitation in 1648 ejected the ardently Royalist dons of Oxford's Corpus Christi College, the bees so esteemed for their long tenancy in the Mellifluous Doctor Ludovicus Vives' quarters also moved out. Supposedly in feminine sympathy with the deposed Charles, they came shortly to nothing. Another colony, perhaps Roundhead, soon settled in the cloister but they too expired after the Stuart Restoration.

Moses Rusden published his *Full Discovery of Bees* while employed as Charles II's beekeeper. Fancifully, he contrasts lion and eagle tyrannies with the warm bonds between the King Bee and his colony. His Dedication displays a quaint mixture of obsequiousness and a craftsman's self-confidence:

> *To the King's most excellent Majesty — May it please your Majesty. I am very conscious that such as presume to make Dedications to your Majesty ought to treat of Subjects lofty and learned, and with a pen eloquent and subtile. But although the subject of this small Offering be mean and the stile be humble. Yet, it speaks knowingly and by experience of Kings, and Chieftains, of War, and Peace; of Obedience and Subjection, of Government and Discipline, of Ingenuity and Labour and the good effects arising from them: and shews how necessary they are, even in this little well-formed Commonwealth of the Bees; whose Being and Prosperity depend upon their Prince, and common Parent. Fathers being the first Rulers, Lawgivers and Monarchs, before Families did swell into a larger Kingdom.*

26 Thomas Mouffet (1658)
The Theater of Insects, p892.

27 Purchas, p33.

With more extravagant praise of King Bees, he hopes to influence those with a malicious eye on Kingly Government. With instincts better than human reason, his subjects love his arbitrary justice and rightly brag *Est Deus in Nobis* (God dwells within us). A humane society needs their eternal laws of 'Prudence, Foresight, Courage, Oeconomy, Fidelity, Regard to Interest, and all those Offices and Virtues'.

While it suited Rusden's purpose to insist on an apian king, under Queen Anne Joseph Warder could conveniently dismiss that 'ridiculous notion'. Naming his book *The True Amazons or The Monarchy of Bees*, he writes a surpassingly grovelling address:

> To the QUEEN's Most Excellent Majesty.
>
> MADAM,
> THERE is nothing can excuse the Presumption I am guilty of in thus approaching your Sacred Hands with so mean a trifle, but the Subject here treated of, which is of Princes and Potentates, Kingdoms and Territories, Prerogative and Property, Dominion and Loyalty, War and Peace.
> I have with a studious Delight, for above Twenty Years past, convers'd with these Innocent Creature the **Bees**, and have not failed... to inform myself, by the most curious Observations, of their Nature and Oeconomy; wherein I find so many, many Things that resemble Your Majesty's happy State and Government.

Continuing in like vein under his Queen's favourable beams, he sadly notes unquiet spirits in England. Nevertheless, by the close of the book she is dead and, without missing a beat, he exclaims:

> Oh, that all the Thousands of this Britannick–Israel were but so loyal to our most gracious King George.

Next came the political and economic verse satire *The Fable of the Bees* by a Dutch doctor settled in London, Bernard de Mandeville. His work opens with the necessary bow to the powers that be, in this case the 'limited Monarchy' of George I. Here 'a spacious hive' prospered, providing a great Nursery of Sciences and Industry.

> They were not Slaves to Tyranny,
> Nor ruled by wild Democracy;
> But Kings, that could not wrong, because
> Their Power was circumscrib'd by Laws.

Under successive Georges, several editions of Thorley's *Melisselogia* commend the ever restive Irish to submit like bees to their rightful sovereign and show 'Courage and Resolution in Defence of civil Liberties and reformed Religion'! With sovereignty more absolute than that of a sultan and all his janissaries, bees gladly obey their kind queen.

Although Swammerdam had already demolished such anthropomorphic perspectives, his ideas travelled slowly and the metaphors continue to flourish: and so do the fulsome dedications.[28]

Written during France's revolutionary convulsions, Jacques Vanière's *The Bees* seems thoroughly ambivalent. On the one hand he presents wild bees as an anarchistic, destructive force proclaiming 'the Rights of Man' while a new queen in

28 For instance, in Thomas Wildman (1768) *Treatise on the Management of Bees.*

the hive establishes order. Yet his last Canto sails to Paraguay, to depict a co-operative bee-like paradise:

> *Their fields and pastures know no separate bounds,*
> *And no litigious fences mark the grounds...*
> *All lies in common; what their crops produce,*
> *Is stor'd in magazines for public use.*
> *All have their province in the general toil...*
> *And grim-eyed war sleeps in his iron cave...*[29]

AN ECONOMIC COMMUNITY

Much of Mandeville's satire[30] has enduring truth, detailing how all sorts of base self-seekers, from unskilled doctors to cowardly generals and cheating tradesmen, make essential contributions to the national hive. Courts punish only 'the Desp'rate and the Poor' while governments and public servants call their 'slipp'ry Tricks a Perquisite'. Extravagance provides work for service industries and dishonesty keeps gaolers busy. Without all these wheels of trade, the hive runs down and provides easy pickings for external predators. A country's prosperity and ease is incompatible with virtue and innocence in politics and economics, so wonderfully greedy people should stop complaining. The State is a Paradise, whose jarring harmonies make music of the whole.

Published in 1769, John Gay's fable *The Degenerate Bees*[31] is another satire. Known as 'honest Gay', 'gentle Gay', he mocks the high society that supports him above his station and doubts his wisdom in siding with Jonathan Swift, notorious ruffler of blockhead periwigs. Even if scoffed at, he declares himself:

> *A stubborn Bee, among the swarm,*
> *With honest indignation warm.*[32]

Bees buzz through eighteenth century letters to editors, rife with class politics. A cynic suggests militiamen should, like bees smoked out to ensure trouble-free access to their honey, be driven from their present situations into enrolling as regulars![33] Bankrupts are 'the most crying grievance of the nation', drones devouring honest workers' honey, writes a self-styled conscientious businessman to the *London Chronicle* in 1759.

But having read *Smellie's Philosophy of Natural History*, another writer[34] commends the insects' mutual regard; a honey-loaded bee feeding a hungry forager is an amiable example to richer people. At a time when poverty was driving emigration, the First Lord of the Admiralty earned this tribute after commissioning naval dockyards to repair our island's floating bulwarks:

> *... giving back some honey to the bee,*
> *Shews justice may with policy agree; ...*[35]

29 Murphy's translation, p55.

30 B. de Mandeville (1714) *The Fable of the Bees, or Private Vices, Publick Benefits.*

31 Most famous as the author of *The Beggar's Opera.*

32 Gay, *The Works of Mr John Gay in 4 Volumes*, Vol 3 p200.

33 *Morning Chronicle* 10.6.1790.

34 *World* 3.12.1790.

35 *Morning Chronicle and London Advertiser* 22.7.1772.

Profoundly distrusting Parliament and finance, one 'Cui Bono' strikes a resonant note. He calls a meeting at a Reading pub to propose the formation of county committees whose own manly exertions would develop the local economy and secure the nation's abundant revenue:

> *Every bee must labour to increase and preserve the little honey which is left, for if the spiteful wasps or hovering drones, should ever more possess the hive, the constitution is stung beyond redemption.* [36]

Scottish Nationalism tinges a letter resenting England's drones' reliance on phalanxes of Scots to fill the professions with their honey. Since every Bee carries a Sting, English taxes should pay Scotland for them. [37]

THE PIONEERING BEE

British colonists freely identified themselves with the hard-working insect. The Pilgrim Fathers soon requisitioned the old country for bees and William Penn, for instance, sought North European Protestants to 'hive off' new settlements in his territory. Within two centuries this crafty, industrious 'white man's fly' reached California. Seen by indigenous Indians as a signal of doom, the hive was a popular symbol of white values.

A skep's thirteen woven rings reflected the continent's first political identity and adorned their paper money. A hive is still the commonest state motif, chosen by almost one third of the Union. The Mormon State of Utah is built around this image, sometimes flanked by sturdy pioneers gazing towards a magnificent heavenly city. Its polygamous founding father, Brigham Young, named his home The Beehive and their legends recount a people known as Jaredites,[38] who reached central America after migrating with their bees across the Ancient World. Andrew Gough's *Arcadia* website links Young's desire to name his state 'Deseret', meaning honeybee, to kindred ancient Middle Eastern words.

One inscription on the George Washington Monument in the US capital pays homage to the Lord Deseret. It may have no greater significance to the national identity than the others but Washington himself was a committed Freemason, to whom bee symbolism is important. Meanwhile, black slaves extended the old folk song:

> *The Lord made the bees,*
> *The bees made the honey,*
> *The Lord made man*
> *And Man made money.*
>
> *De olde bee make de honey comb*
> *De young bee make de honey,*
> *De nigger make de cotton and corn*
> *An' de white folks make de money.* [39]

36 *Morning Post and Daily Advertiser*, October 1784.

37 In a letter to the *Public Advertiser* in 1782.

38 Their provenance is apparently doubted by historians.

39 W.F. Allen et al (1867) *Slave Songs of the United States*.

THE BRIEFLY REVOLUTIONARY BEE

While British beekeepers sang the restored monarchy's praises, republican sentiments were rising across the Channel. In 1740, Jean-Baptiste Simon's *The Admirable Government, or the Republic of Bees* appeared. Four years later, Bazin presented the hive as a republican commonwealth, not an obsequiously obedient society; he credited it with an impressive range of qualities – ancient Persia's commitment to education, Roman patriotism, Scythian frugality, Chinese modesty, Swiss frankness and sincerity, and Turks' sobriety. 'In a word… Almost a treatise of morality.'

After the French Revolution, Jacobins viewed the cell's hexagon, an echo of the country's shape, as a symbol of unbending civic solidarity but the Assembly rejected official use of hive emblems – because the hive has a queen!

Napoleon appropriated the Roman eagle and the Carolingian Hand of Justice and Sceptre of Authority, in addition to Childeric's bees. Such symbolism was everywhere when he crowned himself Emperor in 1804.[40] Nôtre Dame was decked out with golden bees and they were embroidered plentifully on his and Josephine's sumptuous robes. His robe forms the canopy framing the national crest of his First Empire, to appear again on the arms of Napoleon III's Second Empire. Bee motifs proliferated on ceremonial robes and furnishings and in civic coats of arms, such as those of the port he founded at Wimereux. The insects retained their potent symbolism under the restored Bourbon monarchy.

Periodic uprisings racked Europe in the nineteenth century. Living through spasmodic assaults on the Russian autocracy, Nikolai Nekrasov imagined a people unchained and 'countless swarms' building a new world of science and boundless enterprise.[41]

40 Ironically, scholars are not certain that Childeric's insects are bees; some think they may be cicadas, symbols of death and resurrection. Obviously bees would carry more positive symbolism for the self-appointed Emperor, and bees are unambiguous on many public devices dating from the Napoleonic era.

41 Quoted by Juan Antonio Ramirez (2000) *The Beehive Metaphor: From Gaudi to Corbusier*, p23.

(Right) **Fig. 6-4** : Napoleon in Coronation Robes, by Jacques-Louis David
© *RMN-Grand Palais (Musée du Louvre) / Daniel Arnaudet /Gérard Blot*

Fig. 6-5 : Napoleon's Imperial Coat of Arms. Courtesy of Rufus Reade, from previous files by Katepanomegas, Spedona and Blason Roi de Rome.svg
Licensed by Wikipedia Creative Commons ShareAlike

Fig. 6-6 : Civic arms of the port of Wimereux, France
Courtesy of the Mairie

Fig. 6-7 : George Cruikshank's Victorian cartoon depicting British society as a beehive. Victoria and Albert Museum, London
© V&A Images

42 Mary Collier (1739)
The Woman's Labour.

43 A character in
Aristophanes' *The Wasps.*

Britain was not entirely exempt. Mary Collier had complained in 1739[42] that sordid owners garnered the honey of women's labour. The French Revolution inspired the Scottish Utopian James Bonner to raise his voice against the British monarchy, declaring the hive a model enlightened commonwealth. Shelley called for rebellion against aristocratic drones after the miseries of the Napoleonic Wars; all should share Anticleon's ancient resentment against an authority that exacts tributes from virtual slaves in order to butter up friends with honey, cushions, coronets and every conceivable luxury.[43]

Mary Alcock, more in tune with the literate classes, feared a previously contented and orderly hive degenerating and crying:

> *Seize, seize the honey, and lay waste the comb!*
> *Destroy each cell, for labour now is o'er,*
> *We'll feast and revel on the public store.* [44]

Circumspect political reforms quashed incipient revolutionary zeal in Britain. Victoria, the Queen Bee monarch, was idealised. Thomas Nutt approves the 'Tory loyalty' of a hive.[45] Originally drawn in 1840, Cruikshank's Beehive depicts a benign queen ruling over an orderly and productive society; republished in 1867 as 'A Penny Political Picture for the People' by 'their old friend George Cruikshank', it unsuccessfully opposed 'the folly' of a wider franchise.

Meanwhile trainloads from all over Britain attended the 1851 Great Exhibition, that huge showcase of Victorian enterprise. Dominating page after page throughout the summer, the *Illustrated London News* hymned this 'great gathering of industrious bees':

> *…more than two hundred thousand little labourers are diligently engaged in their various daily duties, while their reigning sovereign reposes quietly in her regal apartment, attended to by her subjects with the utmost regard to her comfort and convenience.*

44 Alcock, Mary *Poems* (1799) 'The Hive of Bees; a Fable written in December 1792'.

45 Quoted by J.F.M. Clark (2009) *Bugs and the Victorians*, p70.

46 Clark, p69.

The Industrious Bee

In parallel, during this century of rapid industrialisation the bees' virtuous commonwealth proves a flexible symbol for municipalities, craft guilds and co-operative societies. Clark suggests that the principle of Nutt's 'rational hive', which allowed harvesting without destroying the workers, should appeal both to the owners of teeming factories and to technocratic Whig politicians seeking to manage chaotic urbanisation.[46] Certainly the hive's unceasing labour appealed to the burghers of the new cities and found expression in many civic and industrial badges.

Fig. 6-8 : Coat of Arms of Bacup, Lancs
Courtesy of David Firmstone

Manchester was the biggest of England's congealing mass of cotton towns and bees feature in the civic Arms granted in 1842, are blazoned widely around the magnificent Town Hall and also decorate street furniture.

Fig. 6-9 : Detail of mosaic floor in Manchester Town Hall

Photo by the author

COTTONOPOLIS[47]

For centuries handlooms stood
In cottage corners on damp hillsides.
Then parades of windows lighted upper floors,
Whole families worked on wool and silks
Till cotton came, stuff bleached in fields, stretched
On tenterhooks across those hills,

47 A nickname for Manchester.

Packhorsed to a market hall, each piece
A journeyman's independent livelihood.
But no Ned Ludd could vanquish coal and steam.
Fly shuttle and spinning jenny doomed their craft
And 'hands' swarmed to the whistle's behest.

The shuttle's ceaseless to and fro echoes
The gilded errands of foraging bees.
Cotton bales from around the globe
Became yarn and cloth; workers learned
A language of lips and signs as precise
And intricate as dances in the hive.

The Scout and Guide movements were launched in the early twentieth century, offering ideals of healthy activity and good citizenship, but founder Lord Baden-Powell disturbingly commends the hive for killing their unemployed![48]

Co-operative or Grateful Bees?

Neither aristocracy nor cotton barons could hold the motif all to themselves. Marx's *Das Kapital* points out that humans are not bees, mere means of production; they possess their own imaginative aspirations. The hive as commonwealth nevertheless appealed to working class interests. Droylsden Co-operative Society adorned its grand frontage with a hive and the emerging Trade Unions named their newspaper *The Beehive*. In February 1874 that paper featured an ongoing story contrasting the imagery of a benevolent Queen Bee with the reality of poor wages paid to Victoria's own labourers. When called to supply a good example, a Palace spokesman replied that their wages, augmented by charity, were acceptable; the paper indignantly riposted that bounty is no substitute for a fair wage and allows shabby employers to do great injury to their workers.

But in a tendentious homily Paley argues that beelike busyness and providence would prevent class envy; he even suggests the idle rich do not eat more and deserve pity because they are exceedingly oppressed by want of purpose![49]

And so the different theories of bee society – hierarchy or commonwealth – continue to clash at intervals, interrupted occasionally by most unbeelike dreams of anarchy.

Rebel Bees

At the century's close, while anarchist Communists dreamed of a world 'free of money, masters, the press and property tyrants', a cynic says the spider will always consume the bee unless sympathetic onlookers actually join forces. That the worker alone should win, 'Oh, preposterous thought! Oh, catastrophe!'.[50]

The same symbols still arise in modern political conflicts. In the 1930s slump, workers who would cull leisured drones are presented as envious Marxist revolutionaries crying 'Death to princes. To those who produce belong the fruits'.[51]

48 Robert Baden-Powell (1907) *Scouting for Boys.*

49 William Paley (1832) 'The fable of the beehive' in *Reasons for Contentment Addressed to the Labouring Part of the British Public.*

50 'The Spider and the Bee: a tale for our times', in Louisa S. Bevington, ed. (1895) *Liberty Lyrics.*

51 I am indebted to Claire Preston (2006) Bee, p74-5, for the original untraceable quotation from Chas E. Waterman (1933) *Apiatia: Little Essays on Honey-Makers.*

In 'The Swarm', Australian poet Les Murray likens 'Poor monarchists, clumped round their queen' to half-risen dough.

Written during the recent Troubles in Northern Ireland, Paul Muldoon's satirical 'Bechbretha'[52] imagines a swarm disrupting a garden party of political bigwigs. While some participants hang on to their wine and grab another canapé, Britain slyly captures the bees despite specious stumbling blocks erected by the Irish from ancient precedent:

> *on every conceivable form*
> *of bee dispute,*
> *bee-trespass and bee-compensation.*[53]

DYSTOPIA

As millennia of conflicting meanings collapsed under scientific observation, the poor bees were saddled with a different burden, a generally negative view of the hive's totalitarian nature. In his *Proverbs of Hell*, William Blake says the bee has no time for sorrow. Just so, Crabbe's uncompromisingly efficient and bee-like Widow Goe[54] neglects love or tender cares; yet still death catches her with accounts unfinished. Edwardes depicts a ruthless eugenic state, with the queen a small-brained egg machine, slave to the collective. Even the beekeeper is slave to their behaviour:

> *Nature is always wonderful, but not always admirable... Absolute communism*
> *implies incidental cruelty.*[55]

Preston[56] includes an interesting chapter on the cinema's use of bees, particularly highlighting Leni Riefenstahl's images of Nazi rallies and Fritz Lang's Wellsian *Metropolis* where slave workers obediently perform repetitive tasks underground in tiny, stacked cells. Ratnieks[57] provocatively contrasts hive and supermarket: suckers and foragers are better balanced in the former than cashiers and customers in the latter!

Maeterlinck is ambivalent. Instinct drives bees to make honey regardless of who benefits but our intellectual and spiritual natures equally compel us to create a society.[58] He endorses Pericles' assertion that sufferers in a prospering city are happier than individuals prospering in a decaying State.[59]

METAPHORS MADE CONCRETE

Not only has bee society proved a malleable political metaphor, a spectrum from ideal community to totalitarian horror; it has led to various utopians applying wildly divergent interpretations to actual buildings.[60] Nineteenth century philanthropist Alfred Boucher built *La Ruche de Paris* as a cellular space modelled on a traditional hive design, for a beautiful commune of young artists: '*Mes enfants du bon Dieu! Mes pauvres abeilles!*'[61] ('My children of the good God! My poor bees!') Bruno Taut pictures such a non-possessive utopia as a fecund, joyful place where artists gather nectar and offer honey.

52 From *Collected Poems 1968-1998*. The title refers back to Ireland's ancient laws, cited in Chapter 2.

53 *Poems 1968-1998*, p163.

54 George Crabbe (1807) *The Parish Register: Burials*.

55 E. Edwardes, (1920 reprint), *The Lore of the Honeybee* p.xvii-xviii, xix.

56 Claire Preston (2006) *Bee*, Chapter 10.

57 F.L.W. Ratnieks, 'Are you being served?' in *Beekeepers' Quarterly* (2002) 67, pp26-7.

58 Maurice Maeterlinck (1901) *The Life of the Bee*, p349.

59 Maeterlinck, p344.

60 Juan A. Ramirez (2000) *The Beehive Metaphor: from Gaudi to Le Corbusier.*

61 Boucher, quoted by Jacques Chapiro (1960) *La Ruche*, p31.

Rudolf Steiner argued, among his extraordinary philosophical syntheses, that the hexagon's unique energies not only imprint themselves on the larva and imago but are also embedded in human anatomy.[62] Whatever we may think of that train of thought, he embodied concepts from the hive in two highly regarded *Goetheanum* designs for a University of Spiritual Science. The ascetic Catalan Antonio Gaudi, from childhood delighted by nature, saw his architectural vocation as a collaboration with God. Most famed for *La Sagrada Familia* in Barcelona, extraordinary ornamentation and flowing curves characterise his visionary houses and idealistic co-operative buildings. Catenary arches that mirror a chain of bees and his spires' striking honeycomb structure depart utterly from Gothic or Classical traditions.

In contrast, Paul Scheerbart, dying on hunger strike during the First World War, never realised his dream of a collective paradise on earth without copulation or personal possessions. Built everywhere of glass, it would be 'as if the earth had recovered the precious, lustrous jewels and diamonds'.

Even later, after the overthrow of terrifying Communist and Nazi attempts to build their societies as superorganisms, Modernist Le Corbusier likened cars on a production line to larvae tended by nurse bees and saw houses as machines for living in. His Dom-ino prefabricated apartments slotted and stacked like hives in a beehouse, a hyper-rational principle emerging most famously in his Marseilles block *L'Unité d'Habitation*, a 'harmonious city' with comprehensive communal facilities.

Finally, Mark Thompson created a *Live-in Hive*, a glass cabinet in which his head could live with a colony and his viewers could experience his experience. Akin to Steiner in seeking life's most secret mysteries – a utopia of bees, sun and floral sex organs – his experiment was soon doomed as instinct drove the bees to wrap the intruder in propolis. Kelly describes another of Thompson's efforts at communion with them. He decided to run some distance among a swarm drifting across fences and into a marsh.

The two of them now resembled a superstitious swamp devil, humming, hovering and plowing through the miasma [until] on some signal, the bees accelerated. They unhaloed Mark and left him standing there wet, 'in panting, joyful amazement'.[63]

What an astonishing cargo of conflicting ideals have been loaded onto these small, useful insects.

62 Rudolf Steiner (transl. T. Braatz, 1998) *Bees*, p49ff.

63 Kevin Kelly (2003) *Out of Control: The New Biology of Machines, Social Systems, and the Economic World*, p8.

7 : The Bees' Own City

The fridge hum of the hive belies
Warm lives in a complex world,
One we frame but they possess,
A commonwealth of amazons
Needing no generals to marshal them.

People have always known that honey bees come in three types or 'castes' –
worker, drone and queen – and previous chapters have discussed elaborate
ideas about their wonderful and worthy lives.

Bartolomaeus Anglicus[1] asserts offenders quite properly punish themselves and
die by their own sting. They 'ordain watches after the manner of castles' while others
sleep until the 'morning trump'. Work is honestly shared and not one:

> *... doth espy nor wait to take out of other's travail, neither taketh wrongfully, neither*
> *stealeth meat, but each seeketh and gathereth by his own flight and travail among*
> *herbs and flowers that are good and convenable.*

The Renaissance essayist Montaigne[2] asks whether any human society is better
ordered, consistently maintained and more functionally diversified than the hive.
Butler[3] itemised the inhabitants' prudent government and intelligence: knowing their
own hive, distinguishing friend from alien, recognising when drones are needed or
should be culled. They are clean housekeepers and groom carefully, and the queen
lays appropriate eggs in the right cells. Thorley notices how teams work together to
lift burdens of hive dirt or dead bees.

But these bywords for civic virtue also harbour extreme aggression, singly
powerless as a drop of water but invincible en masse,[4] bears or lions at bay,[5] a
burning bush of heroism,[6] intent only on the future of the hive.

Tufted Coronels and Captains

Elaborate ideas have been expressed about the ranks of worker bees. Like Pliny,
Butler[7] thought there were different degrees of dignity: he notices officers 'marked
with a yellow tassel sometimes curled like an ostrich feather'. Later writers
distinguished ever more ranks. Besides Watchmen, Scout-watches and Sentinels to
repel flying thieves, Levett[8] posits a royal guard whose officers rank from Generals
to Sergeant-Majors. Even more imaginatively, Topsell[9] lists Dukes, Ambassadors,
Orators; also Souldiers, Pipers, Trumpeters, and Horn-Winders!

Purchas curtly dismisses such fancies; those tufted colonels are simply younger
bees whose fur has not yet worn off.[10] But he says they delight to play abroad and
after a long winter indoors will snatch any sunny moment to ease their bodies.
Reaffirming the colony's readiness to rob their neighbours when necessary, among
varied hums he distinguishes the loud, threatening tones of approaching robbers.

1 Bartolomaeus Anglicus
(1240) *On the Properties
of Things*.

2 Michel Montaigne
(1580-88) *Essays*.

3 Charles Butler (1609)
The Feminine Monarchie.

4 Samuel Purchas (1657)
*A Theatre of Politicall Flying
Insects*, p16.

5 John Thorley (1st ed. 1744;
4th 1774) *Melisselogia or the
Female Monarchy*, p16.

6 Maurice Maeterlinck (1901)
The Life of the Bee, p21.

7 Charles Butler (1609) *The
Feminine Monarchie*, Ch 1,10.

8 John Levett (1634)
The Ordering of Bees, p69.

9 Edward Topsell (1658)
*The History of Serpents
or The Second Book of
Living Creatures*, p638.

10 Purchas, p16.

Fig. 7-1 : Worker bee feeding
Courtesy of Paul Embden

Watching his observation hive, Réaumur concludes the occupants comprise a sexless commonwealth under a despotic queen.[11] Maeterlinck finds these seething 'huddled raisins' incoherent and incomprehensible. On the other hand, Edwardes fancies he sees:

> *... something curiously human-like...incessant hurryings...; chance meetings of friends on street corners; altercations where we can almost hear the surly complaint and tart reply; busy masons and tillers and warehouse-hands...*[12]

The nineteenth century German Johannes Mehring deems the colony equivalent to a vertebrate, in modern terms a single superorganism. At the summer peak, 50,000 workers maintain and digest while the queen and drones supply genitals. Professor Jürgen Tautz of Wurzburg[13] even calls it a mammal, sharing mammalian characteristics – similar temperature, few queens, feeding via glands, womb-like cells and a mental capacity eclipsing some vertebrates. The community also solves survival problems – varied environment, reliable food and safe housing – through division of labour.

DIVISION OF LABOUR AMONG WORKERS

Aristotle[14] was the first to record the workers' different jobs, starting three days after hatching. Indoors there are cleaners and comb-builders, others secure cracks with propolis and restrict the entrances, and yet others pack honey into the cells or tend

11 René-Antoine de Réaumur (1741) *Notes to serve for a history of insects* 9th and 10th Memoires.

12 Edward Edwardes (1908) *The Lore of the Honeybee*, p71.

13 Jürgen Tautz (2008) *The Buzz about Bees*, Prologue.

14 *History of Animals Book VIII (IX)*, transl. Balme, p335ff.

the father (sic) and brood. Outside duties include water-gathering as well as foraging for nectar and pollen. He records the fastidious removal of their dead and of rubbish, relieving themselves outside and, more questionably, their hatred of unpleasant smells and perfumes. Thorley establishes that every worker undertakes the full range of duties: gathering wax,[15] fetching honey, keeping guard, carrying the dead, cleaning the hive and killing drones.

Those tasks have now been roughly sequenced in each individual's five-week life, starting with cell cleaning, followed by a mix of indoor jobs, finally graduating to foraging. However, what remains unclear is how far the sequence is determined by an individual's genes or adaptability to the hive's fluctuating needs.

House bees

After the orgies and excitement of a departing swarm, Maeterlinck[16] says a languid, faithful nucleus stays on, tidying up and resuming their nurture of 60-80,000 nymphs in sealed cells. Immediately a cell cap breaks and eyes and antennae appear, nurses run to help, cleaning and brushing the newborn and giving her honey. They start to fan other cells. Others stand by the entrance to fan and cool the hive. Swammerdam[17] sees them all as slaves, feeding the young, packing winter stores in vacated cells and even destroying some of the brood if larder space is short.

Building their strength on a pollen diet, apprentices emerge for occasional aerobatic practice, returning like toddlers for frequent rests. Two days on, they start exuding wax, capping cells. Longer flights and spells on guard develop their repertoire. At three weeks old they learn the dance.

Malformed bees are ruthlessly ejected, says Warder,[18] and the dead solemnly disposed of. According to Vanière, unless too busy foraging, several bees will carry a body up to thirty yards.

> *With the cold corpse a melancholy train*
> *In slow procession seeks the neighb'ring plain*
> *And the last rites and fun'ral honours pay.*[19]

Builders

Only young bees produce wax, losing the organs as they age. A special tool on the hind leg draws wax-scales from their pockets,[20] and astonishingly they can two-thirds fill an empty hive with comb within a week.[21]

Maeterlinck[22] vividly describes the work, their most urgent task in a new home being to provide brood cells for the queen already clamorously dropping eggs on the floor. Workers climb in solid columns and then, gripping each other by the front legs, form long chains and garlands. Eventually they hang in a cone from the apex while eight white and transparent scales strangely sweat from the youngest bees' four abdominal pockets. One bee butts her way to the summit and kneads the first keystone with her mouth and claws. After others add more wax, a sculptor arrives to shape the cell. More build on around it and a two-sided tongue grows down. Another founder bee places another keystone a nice distance away, and so on until

15 Thorley, p167. It was not yet known that young bees exude the wax from their bodies.

16 Maeterlinck, p62.

17 Jan Swammerdam (1758) *The Book of Nature*, p190.

18 Joseph Warder (1712) *The True Amazons or The Monarchy of Bees*, p20.

19 Arthur Murphy (transl. 1799) Jacques Vanière *The Bees: a Poem*, Canto 1.

20 Edwardes, p201.

21 Thorley, p126.

22 Maeterlinck, pp111-12, 175-6.

the dimpled, back-to-back frames of hexagons take shape. Eventually there will be four sorts of cells: for worker brood, larger for drones, for stores, and finally a few thumb-like queen cells.

GUARDS

A later phase of house duties is to guard the hive against robbers or beggars. Alert on the threshold, their front legs are ready to grasp intruders. Raiders looking to slip in are betrayed by their alien smell and hovering, inquisitive flight. Purchas[23] advises prudent beekeepers to help the guards with narrowed doors in late July but, in the event of battle, damaging the robbers' home combs will distract them. Either measure is more effective than Virgil's practice of dowsing them with dust or smoke and water. Scandalised by their belligerence, Thorley excuses this 'martial, unpeaceable Spirit, and notorious Injustice' as arising from their 'perfect Abhorrence of Sloth and Idleness, [and] insatiable Thrift'.[24]

Fig. 7-2 : Guard bees on hive alighting board
Courtesy of Paul Embden

SCOUTS

Besides occasionally seeking a new home for a swarm, their regular job is to locate good foraging grounds and dance its whereabouts.

> *SCOUTS*
>
> *Sunny days with barely a breeze*
> *And daphne scents first fill the air;*
> *Saffron accents the crocus' open face*
> *And pussy willows' sleekness turns*
> *To ragged gold. Early spring patrols*
> *Bring samples, gyrate like disco dancers*
> *Self-absorbed but mapping paths for multitudes.*

23 Purchas, p114.

24 Thorley, p163.

Then, as the seasons roll, they locate a succession of farm crops, tree blossoms and garden flowers until brooding ivy's last dark gift.

FORAGERS — 'LONG SEARCH, TRUE JUDGEMENT, AND DISTINCTION OF THINGS'[25]

Fig. 7-3 :

Drawing nectar from marjoram

Courtesy of Paul Embden

A week after hatching, a young bee takes her first cleansing flight, expanding as she flies. After another few days of anxious orientation she starts foraging, gradually broadening her terrain, says Jonathan Swift, as a discriminating reader should. It is now known that the average worker flies 900 kilometres in her few short weeks while a colony would ring the world ten times each day.

Butler[26] watched them setting out keenly in the morning sun and returning as loudly as merry gossips, laden with ambrosia. His attentive eye gave us the first record of how they work. Fangs gather pollen, which forelegs nimbly convey to hindlegs. Their tongues gather nectar and swallow it down into bladder-like bottles until their 'little bellis strut with all'. Back at the hive they unload onto house bees.

So eager are they to get to work, each one beats her head against living walls, says Maeterlinck. He believed nectar and pollen days alternate; in fact availability and the hive's needs govern collecting patterns. Others have observed how they catnap in a vacant cell before combing their heads and setting out again. Possibly over

25 Jonathan Swift (1704)
The Battle of the Books.

26 Butler, Ch 6, para 44.

Fig. 7-4 : Checking a returning forager
Courtesy of Paul Embden

seventy percent of their day is rest time. Edwardes[27] is characteristically lyrical about their commuter patterns, early risers lancing straight into the sunshine on repeat excursions while laggards make little impetuous rushes to the step before pausing to flutter their wings and adjust to the light. Meanwhile, homing bees sail in sedately like laden bronze argosies, perhaps resting on a nearby bush or falling exhausted into the grass before entering.

Like the Romans, Butler still measures the foraging season between the rise and setting of the Pleiades. Depending on the weather (and snowdrops), we may see them in mid-January. By late August the flowering season has peaked and combs should be loaded. Then, says Thorley,[28] the beekeeper can take his share but must leave a judicious amount and top up later with winter syrup.

THE QUEEN

Without a doubt of her sex, Xenophon saw her job as household management:

> *I think the queen bee is busy about just such other tasks appointed by God… She stays in the hive and does not suffer the bees to be idle; but those whose duty it is to work outside she sends forth to their work; and whatever each of them brings in she knows and receives it, and keeps it until it is wanted. And when the time is come to use it, she portions out the just share to each. She likewise presides over the weaving of the combs in the hive, that they may be well and quickly woven, and cares for the brood of little ones that it be duly reared up. And when the young bees have been duly reared and are fit for work, she sends them forth to found a colony, with a leader to guide the young adventurers.*[29]

Also having no dispute about her sex, Chaucer declares the queen is elected for stingless mildness:

> *Thise flyes that men clepeth bees, whan they maken hir king, they chesen oon that hath no prikke.*[30]

27 Edwardes, p54-5.

28 Thorley, p151.

29 *Economics VII*, 32-34.

30 *Canterbury Tales; Parson's Tale*, p468.

Fig. 7-5 : Marked queen with attendants

Courtesy of Paul Embden

Not elected, declares Purchas,[31] because 'the vulgar often want judgement', and it is another mere fable that a royal couple rule together from a fenced palace high in the hive.

Despite knowing the facts, Topsell insists on describing a king – tall, personable, heroical and bound to the colony by oath. Slow to punish, he will nevertheless do so:

> *… if he finde any of his sons to be a fool, unhandsome, that none can take pleasure in, rugged, rough, soon angry, fanciful or too testy, ill-shaped, not beautiful or Gentleman-like…*[32]

Generally the subjects willingly and lovingly submit but maintain a certain prerogative in voicing their opinions and destroy anti-social elements 'by a common consent and by a Parliamentary authority… for fear the whole Swarm should be divided and distracted into many mindes…'

Fancifully, Rusden draws an analogy between the hive and lion or eagle societies, each a commonwealth of cheerfully obedient servants under a benign master. Attendants will loyally die around him rather than desert a king fallen outside the hive.[33]

It was well-known that a colony is placid under one monarch, pines and dies without one but is restless and useless if rivals emerge. Rusden asserts that, unless they swarm or the beekeeper manages a merger, the king rallies his troops to destroy princes. He walks among them like a general, but (in a less than heroic account) hastens from troubled times with his attendants via chain ladders of guards to his throne at the top of the hive.

In fact, it is the emerging usurpers who fight and then, not for conscious authority (concludes cool, scientific Swammerdam), but simply because universal ruin would

31 Purchas, p16.

32 Topsell, p639.

33 Moses Rusden (2nd ed.1685) *A Full Discovery of Bees*, p17.

follow two laying in the same cells. Eggs are laid in order: workers, a few queens, then drones near swarm time. He bleakly remarks that workers:

> *… like women virgins past child-bearing, serve only [to] labour in the economy of the whole body… nourishing and educating.* [34]

Given her two duties, to lay eggs without respite and to rule, Thorley debates why a female is given royal dignities. Competing drones might destroy the State – 'an Aristocratical Government would be utterly inconsistent with its Safety' – but he is well aware that two days of civil war can ensue between rival queens.

When he experimentally hid the monarch, Rusden observes how some bees crept while others ran up and down like men in a tumult. But once located, 'their actions and voices proclaim it at the door'; search parties return and 'all together seemed to make a cheerful noise among them'. Thorley describes the same confusion, lamentation and transports of joy. [35]

Claiming a truly new observation that again shows Nature's subtlety, Huber[36] notes queen larvae's imperfect cocoons, which ease the job of slaying excess embryos before strife arises. Maeterlinck[37] thinks differently, saying the workers, having decided to nurture future queens with additional royal jelly,

Fig. 7-6 : Young queens in deadly combat

Courtesy of Paul Embden

then fiercely guard these few royal embryos against a newborn rival's attacks. She, exhausted, will retreat for several days of angry war song while others are detained in strengthened fortresses. When finally born, these are strong and able to lead a swarm, and the elder queen too is ready to lead the second swarm ('cast').

The modern Butler[38] identifies a 'queen substance' secreted to ensure the colony's cohesion. A newborn virgin queen will wander round damaging immature queen cells, piping and listening. Workers seem indifferent to her until replies come from other emergent queens; only then will they engage in the ensuing fight.

Despite all past grand imaginings and discomfiture about her gender, the queen is now recognised as a mere egg-laying machine with a fertilised reservoir and a fund of drone eggs she can select for their larger cells. She enjoys one or more brief mating flights and then, for a few years of fecundity, works and lives as a lavishly attended slave within the hive.

Drones, 'greedy lozells'

Butler[39] knew they were male, obscurely necessary for breeding, and perhaps helping to warm the hive. Again forestalling any feminist conclusion, he notes drones' louder voices but quietness becomes the female! Perhaps they are 'not altogether idle', building the king's house, warming brood or serving as butler or porter, maybe having a reproductive function, suggests Moffett in his *Insectorum* (1658).

Their physique – small-brained and anatomically unfitted for foraging – explains their indolence. Occasional modern writers suggest they have other uses,[40] such as warming the hive, maybe cleaning, secreting an albuminoid-rich jelly important to brood.

34 Swammerdam, p169.

35 Thorley.

36 F. Huber (2nd ed. 1808) *New Observations on the Natural History of Bees*, pp94-5.

37 Maeterlinck, p204ff.

38 C.G. Butler (1954) *The World of the Honeybee*, p101.

39 Butler (1609), Ch 4,1.

40 Jeremy Evans (1989), *Complete Guide to Beekeeping. BBJ* 29.7.1920. Annual report of Smithsonian Institute 1909, quoted by Herbert Brown (1923), *A Bee Melody*, pp162-3.

Fig. 7-7 :

A drone. Note its blunt body and huge eyes

Courtesy of Paul Embden

Nemesis comes in August, when the queen has abundantly displayed her fertility, and has been called 'Tom Drum's entertainment'.[41] Butler[42] says laconically that the workers 'wax weary of their mates and… like their room better than their company'. Beekeepers should cue tardy workers to the job if their brothers resist eviction once breeding is done.[43]

Other accounts of the slaughter are horrific. After months of cherishing them, says Swammerdam, the bees are suddenly 'inflamed with so much hatred… that they unmercifully and for no crime kill them'.[44] In his glass hive, Huber[45] saw workers tear off drones' antennae, legs and wings, then repeatedly sting their bellies, even after they seemed dead. Next day, refugees from other hives would be furiously destroyed. During these long days of carnage, their brood cells too are razed and, says Bazin, 'the hive is cleansed… as it would be from a contagion'.[46]

Maeterlinck[47] observes workers emerging from many hives on a single day with 'cold, deliberate indignation' to mob the drones, even sawing them in half. Skulking wounded survivors will be starved to death while escapers will be forbidden re-entry and die of cold overnight. Only the poorest hives, desperate for a pregnant queen, delay this massacre and thereby doom all to die of poverty. The creatures' 'deep, vibrant, horror-laden note'[48] reiterates the apocalyptic atmosphere.

Brown paints a different picture, the killing as gentle as any Chinese gentleman's obedient suicide. For some days the workers deny him rations, then take him to the hive door and by little jerks move him:

> … *down the alighting board, a small portion of the wing bitten away, and a first attempt made to fling him overboard. The executioners seldom finished their work with one trip; the drone with enfeebled wing takes another flight and yet another, until at last, as a willing martyr to the race god, he resigns himself to his fate…*[49]

41 Forcible ejection of troublemakers, according to *London Literary Gazette* 5th Feb 1825.

42 Butler (1609), Ch 4, 14.

43 Purchas, p41.

44 Swammerdam, p167.

45 Huber (1841) p82.

46 Gilles Bazin (1744) *The Natural History of Bees*, p189.

47 Maeterlinck, p287.

48 Edwardes, p245.

49 Herbert Brown (1923) *A Bee Melody*, p160.

Von Frisch says weeks of slowly rising hostility precede an inglorious end. Lacking stings and too stupid to find their own food, drones obstinately resist eviction or try to blunder back. The modern Butler[50] portrays a lower key battle, where they are simply driven to the comb edge to die of cold and starvation and are only stung to death if still struggling to escape. Longgood,[51] however, strikes the melodramatic note again, referring to the great rattling noise of tormented hulks and the workers' obvious glee in the carnival of death.

BROOD

At the midsummer peak of laying there will be around 40,000 larvae at different stages of maturation, with the queen refilling cells as adults hatch, but in winter none. Arranged in concentric rings on adjacent combs the brood forms a ball overall, easier to tend and maintain in a constant climate.

In a bath of diluted pollen ('bee-bread'), the larvae are tended by nurses until the cell is sealed and imagos form. Workers bite their way out after twenty-one days.

This chapter closes with two esoteric notes. Metamorphosis was a mystery until modern times and, according to Thorley, constitutes an 'Almost lively Emblem or Image of the Resurrection'![52] He fancies that queen eggs in their mansions are golden and bypass the grub stage to emerge as perfect adults. Steiner[53] relates the larvae's different maturational periods to cosmic influences. The queen's sixteen larval days keep her a fertile Sun child; a drone's twenty-four days subject him to the greatest Earth influence and account for both his fertility and high flying strength; the worker's twenty-one days measure and incorporate the Sun's rotational cycle.

Such are the imaginative faculties humans focus on these intriguing insects.

50 C.G. Butler (1954), p69.

51 William Longgood (1985) *The Queen Must Die*, p169-70.

52 Thorley, p117.

53 Rudolf Steiner (1998 ed.) *Bees*, p103-4.

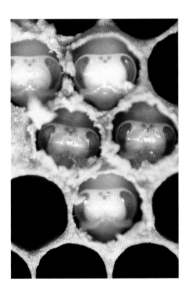

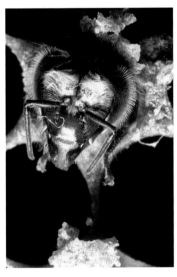

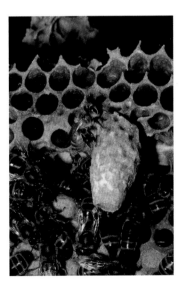

Fig. 7-8 : Worker bee pupae, uncapped for photo

Courtesy of Paul Embden

Fig. 7-9 : Worker bee biting her way out

Courtesy of Paul Embden

Fig. 7-10 : Worker bees tending a queen larva in a still uncapped cell

Courtesy of Paul Embden

8 : Awesome Swarms

Virgil and many since have described this three-act drama, from the mysterious sound signals that presage swarming, until thousands suddenly ascend like a cloud of smoke. There may be a battle before one queen's army masses placidly on any convenient prop while the scouts locate a future home. A skilful beekeeper has prepared one for them to march into – but sometimes they have other plans.

Opportunistic Lodgers

Programmed to find shelter wherever chance presents, bees may choose any one of a marvellous variety of homes. Jonstonus[1] retells an ancient account of a swarm creeping into a sepulchre and 'making abundance of combs in the dead bodies' of two sisters. When a thunderstorm later broke open the tomb, everything was miraculously unharmed. Building repairs in Corpus Christi, Oxford, revealed a century-old nest with a vast quantity of honey above the Professor of Rhetoric's study, neatly linking honey, oratory and the college's reputation as a bee garden of sweet learning. In somewhat poignant contrast, a poor household was beset by inept workmen opening a wall and removing roof tiles in search of a swarm; finally, a fire in the hearth quickly solved the mystery when with 'a strange uproar and confusion… considerable Numbers came down into the Room, thro' the fire and the Smoke'.[2]

1 Johannes Jonstonus (1657) *A History of Wonderful Things*, p245.

2 John Thorley (1744) *Melisselogia*, p149.

Fig. 8-1 : Beekeepers and a wheelie bin to the rescue
Courtesy of Heather Leonard

Some perished, some left by an opened window but most escaped via an old passage to settle outside. One must hope the runaways were then caught and brought some recompense for the workmen's damage! Curious resting places fascinate the Press. Jam or sweet factories may seem a godsend[3] but there is surely less reward in settling in a pottery kiln or plane wing, stopping a church clock, or smothering a buoy in Southampton Water.[4] The scene in Fig. 8-1 would have surprised shoppers heading for the nearby supermarket.

Even people may find themselves acting as a suitable sofa. A swarm inside his terrified maid Anne's veil presented a most affecting spectacle, says Thorley, but he 'entreated her… to stand her Ground and keep her present Posture', while he removed two queens. Thereupon, without a single sting, the throng followed to a waiting hive and Anne's 'gallant Behaviour… immediately formed her into a perfect Heroine…'[5]

Letters to The Times in August 1967 recounted two coolly constructive responses to such untoward situations. Should you happen unprepared on a swarm in woodland, Irene Lloyd advises shaking the insects down over yourself, then gingerly walking home to brush them off with a long feather. Lady Plowden tells of a gardener, fortunately wearing a cap when a swarm landed on his head, who very slowly went and lay down at the door of an empty hive and the bees quietly walked in.

They have been known to settle and drink from a sleeper's lips. A less curious location but a surprising report of bee behaviour comes from Captain Stedman's Surinam. Lodged for greater security and comfort in a tree house, he was perturbed when a visitor was badly stung; an immense nest of 'Wassee Wassee' had lived unknown in the thatch of his aerial dwelling. The appalled captain ordered the creatures' destruction as he fled but an old neighbour said:

> Massa, they would have stung you long ere now had you been a stranger to them; but they, being your tenants, that is, gradually allowed to build on your premises, they will assuredly know you and yours, and will never hurt either you or them.[6]

Sceptical, the captain tied the fellow up as a hostage, before ordering a naked boy to climb up and test his theory. Though the lad shook the nest and its inhabitants buzzed about everyone's ears, nobody was stung. Thereafter the captain accepted the lodgers as his bodyguard and compensated the old man with a gallon of rum and four shillings for his pains.

OMENS GOOD AND BAD

Swarms were occasionally good portents, bestowing wisdom and oratory and signifying victory in battle. Despite soothsayers' pessimism, Pliny viewed one at Arbalo as a positive pointer to Drufus's victory.[7] John Evelyn records how the king forbade disturbance of a swarm hived in 'his Majesties ship, the Oxford fregat'.[8] Perhaps rationally, since you might be able to hive a swarm, French folk traditions consider a strange one alighting as the harbinger of a prosperous year.

More often, though, a swarm was a bad omen. Scipio delayed his expedition until one in his tent had been propitiated.[9] Rome was in a state of dread when, among other strange phenomena in 64 AD, a throng settled on the Capitol. Emperor

3 British Bee Journal 19.8.1950.

4 The Times 22.7.1950, The Times 11.7.1951, British Clay Worker 10.1972, Beecraft 12.1981.

5 John Thorley (1744) Melisselogia, p153.

6 Included in W.C. Cotton (1842) My Bee Book, pp338-9.

7 Pliny Natural History of Animals, Book XI Ch18; Cicero Of Divination 1,73; Just.Epit. 23, 4, 7.

8 John Evelyn Diaries, June 1662.

9 Livy History of Rome, 21, 46, 2 passim.

Claudius's death was foretold by a swarm in his camp and Pompey's defeat at Pharsalus was augured when one settled on his standard.

In much of Christian Western Europe, says Fife, swarms portended death.[10] For instance, if one settles on a dead branch, the tree will die or the bees fail; if around your house or returning to an old hive, someone will soon die. A queen alighting on a person means immediate death. Such ideas have evident affinities with ancient beliefs about bees and souls but the last is at odds with the notion that a settling swarm bestows eloquence.

ACT ONE: THE CALL TO SWARM — WHEN AND WHY

Swarms arise when the queen is laying prolifically and brood space is crowded. Comb might even be built outside the hive; then, Butler said, a swarm must be provoked or extra space be stitched onto their skep.[11] Modern box hives are much easier to extend.

They choose the middle of a warm day. Traditional rhymes confirm the ideal season, but will the rhymes change if climate change brings earlier swarming?

> *A swarm of bees in May is worth a load of hay/a cow and a calf that day,*
> *A swarm of bees in June is worth a silver spoon,*
> *A swarm of bees in July isn't worth a fly/a butterfly.*

Aristotle[12] writes of the monotonous and peculiar hum that forewarns; Bartolomaeus Anglicus also notes the 'voice of the host, as it were arraying itself to pass out with the king'.[13] But after several watchful years, Alexander de Montfort disagrees, attributing this hoarse accent to a failing and seditious queen 'caressing the new souldery, whom he endeavours to inebriate, and draw into revolt against their sovereign'.[14]

Deeming it treason for the princesses or others to imitate the queen's song, the beekeeper musician Butler even transcribed these notes and their rhythm. He asserted the harmonies provide the ground of the musicians' art and the swarm's 'solemn procession sets a pattern for human pomps' and in 1623 composed 'The Bee Madrigall'.[15]

Professing to know little of their negotiating language, Thorley nevertheless asserts the queen sings the four lower *Cliffs* while the princess humbly petitions the queen-mother for leave to swarm, singing the upper register. After a day or so of silence, she will ask again and then 'the Queen, with a very audible voice, [gives] her Royal Grant, proclaiming it as by Sound of a Trumpet thro' the whole Kingdom'.[16] With an ear close to the hive, he distinctly heard what was acting on 'the Throne', a crescendo as the swarm readies and then, as in human partings:

> *… the greatest lamentation among the Branches of the Royal Family. Notes of Woe expressive of the deepest Sorrow, as tho' they were taking an eternal Farewel of one another. It was really in some Measure moving and affecting.*

The sequence of signals was scientifically established by Gould and Gould[18] in 1988.

10 A.E. Fife (1939) *The Concept of the sacredness of bees, honey and wax in Christian Popular Tradition*, p464.

11 Charles Butler (1609) *The Feminine Monarchie*, Ch 5 paras 21-3.

12 Aristotle *The History of Animals*, Bk 9, Ch 27 para 13.

13 Robert Steele (1905 ed.) *Mediaeval Lore from Bartolomaeus Anglicus*, p124.

14 Quoted by Gilles Bazin (1744) *The Natural History of Bees*, p177, from *Springtime of the Bees*.

15 Inserted in Ch 5 of the 1623 Edition.

16 Thorley, p145.

17 Thorley, p146.

18 J.L. and G.G. Gould (1988) *The Honey Bee*, p53-4.

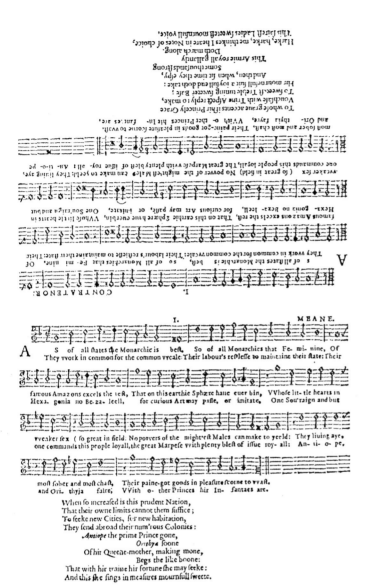

After scout bees' distinctive buzz as they dash through the hive, the reigning queen vibrates her thorax against the comb, producing a specific 'tooting' that excites a queen grub to 'quack'.

ACT TWO: THE SWARM ITSELF

Virgil speaks of the swarming cloud sweeping aloft and darkening the skies.[19] Early writers thought swarms comprised one birth cohort of brood and a young king. Aristotle[20] wonders whether the king is carried. Varro conjures up a family dispatching rebellious adolescents 'as Sabine women used to send out their surplus youth'.[21] The youngsters are equally pleased to go since they despise their elders and their orders!

Painting radically contrasting scenes again, Bartholomaeus Anglicus and Alexander de Montfort concentrate on the monarch's role. Bartholomaeus says:

19 Dryden's *Virgil*, Georgics IV, p108.

20 Aristotle, Book 9 Ch 27 para 6.

21 Marcus Terentius Varro, Book III ch xvi.

When he passeth forth... he is beclipped about with the swarm, as it were with an host of knights...[22]

But Alexander, equally vehement, sees an ugly tyrant possessing the Seven Deadly Sins, a quarrelsome prince departing:

... like a traitor, or piece of counterfeit money, that dares not shew itself. As soon as the sun shine upon his head, his bad qualities appear, and cause one part of his people to revolt.[23]

Fig. 8-3 : A swarm in the air
Courtesy of Paul Embden

22 Bartolomaeus Anglicus (1240) *On the Properties of Things*, Ch 12.

23 Quoted by Bazin, p177, from *Springtime of the Bees*.

24 Butler, Ch 5.

25 Samuel Purchas (1657) *A Theatre of Politicall Flying Insects*, p79-80.

26 Despite Butler's renowned work, several writers continued to call her King.

27 Moses Rusden (1685) *A Full Discovery of Bees*, p25.

28 Thorley, p145.

Butler's fresh insights[24] include proof that both the swarm and those that remain include bees of all ages, plus a queen and drones. But a black cloud will defer their departure, even when already dancing in the air. The best swarms rise before knapweed blossoms but any emerging in blackberry time weaken the hive and will probably die themselves.

Opinions conflict on how the swarming decision is made. Purchas[25] thinks the community decides but Rusden asserts the old king[26] gives the order when he fancies a move from the 'stoping and noisomness' of old combs to 'a more rich, pleasant and flourishing Kingdom'.[27] Thorley takes yet another position; with the queen-mother's consent a young princess leads a cast (secondary swarm) about a week after a prime swarm. But if 'the Royal Grant is withdrawn... all the Royal Issue will be slain'.[28] Normally swarms and casts will not exceed three.

Seeing that young queens hatch fully ready to fly, Swammerdam assumes they immediately drive out the old queen. With workers already singing round the maiden's cell and drones ejaculating wildly to send her (he wrongly thought) 'to be sure, from her joy' to provoke a swarm.[29]

Finally, Huber[30] establishes that the old queen leads the first swarm but only after laying eggs in the several queen cells prepared while she was laying drones. Disburdened, she can fly and new queens are already hatching.

Vanière draws a fanciful scene of her trumpeting, clapping her wings and calling a swarm to revolt. After debate and seditious plots, they follow her command and:

> *The drones, in peace a dull inactive crew,*
> *But ever prone new measures to pursue,*
> *Rush to the gate, the emigrants excite,*
> *And with loud clangour urge them to the flight.*
> *Sublime upon the summit of her hive*
> *The Princess sees her troops in crowds arrive...*
> *And crowds of legions quite obscure the day.*[31]

Little new emerges for nearly 200 years. Citing Swammerdam, Maeterlinck[32] says around thirty new queens may lead an exhausting swarming fever. He adds that the swarming mood can be deflated if the beekeeper destroys developing queen cells.[33] But more usually the third queen is allowed to slaughter further pretenders while vigorous young bees work like Trojans to rebuild their colony. Adjusting to circumstances, when supplies are short they slay royal brood and simply forage but build queen cells when fertility declines. The casts themselves are fraught. An unencumbered young queen may fly too far and lose them all, or she may fail to be fertilised. After twenty days she is irrevocably a virgin, by parthenogenesis producing only drones and causing the colony's rapid collapse.

Besides the acoustic signals mentioned earlier, Gould and Gould further clarify the swarming sequence. In a preliminary vote, those bees in favour of swarming start building new queen cells while those against tear pieces off. Only gradually do the constructors outnumber the destructors. Once a princess grub begins to quack, workers forcibly detain her until needed, by standing on her cell and shaking it vigorously. This stimulates others to forage and stock up while the old queen stops laying. Thus a primary swarm under the old queen is readied for leaving, shortly followed by a cast with the next emerging queen.[34]

Morse and Hooper[35] note that scouts start searching and workers gorge themselves in readiness for the perilous journey. After an early rest nearby, the swarm will fly fast and low. Pheromone signals bind them while scouts dance to indicate the location of their future home.

Fig. 8-4 : A resting swarm
Courtesy of Paul Embden

29 Jan Swammerdam (1758) *The Book of Nature*, p187.

30 François Huber (1808; 1841 ed.) *Observations on the Natural History of Bees*, pp106-7.

31 Arthur Murphy (transl. 1799) Jacques Vanière, *The Bees: a Poem*, from Canto II, p26.

32 Maurice Maeterlinck (1901) *The Life of the Bee*, pp213, 202-4.

33 Maeterlinck, pp46-7.

34 J.L. and G.G. Gould (1998) *The Honey Bee*, p54.

35 Roger Morse and Ted Hooper (1985) *Encyclopaedia of Beekeeping*, pp370, 377.

WARRING SWARMS

Once all tooting and piping has ceased, further queens will dare to hatch and fight for dominance. A colony cannot tolerate two monarchs and the ensuing furious battles are appalling and fascinating. Virgil tells of martial clangours, hoarse alarms and then:

> *The shocking squadrons meet in mortal fight.*
> *Headlong they fall from high, and wounded wound;*
> *And heaps of slaughtered soldiers bite the ground.*
> *Hard hail-stones lie not thicker on the plain,*
> *Nor shaken oaks such showers of acorns rain.* [36]

If a sprinkling of best beer fails to pacify them, Butler describes one such war, which lasted:

> *... ful two days and two nights... wherein such havok was made that the better part of these brave soldiers (a mournful spectacle!) lay, soom ded, soom half-ded, sprauling on the ground. At the last it was my hap to spi one of these queens at the hive-skirts, in a cluster: which taking up, Nou (qot I to one that stood by mee) heer is she for whose sake all this slaughter was made: but an hour after my Soon found the other ded on the ground.* [37]

With both queens slain and most of the swarms, another colony quietly accepted the survivors.

So great is the tension between rivals, they will defy unsuitable weather and even risk their monarch's life. Should he perish, the swarm will return home 'to their relinquisht prince'.[38] Bazin established that a swarm may include two or even three queens and split, but if the smaller party rejoins them in the new hive there will be a fight.[39] We now know that one newly hatched queen will dominate, destroying rival cells or using her sting just this once to fight others to the death. Workers will not directly attack but may ball around and suffocate the rival or simply take a ringside seat.

In *Paradise Lost*, Milton even likens the war between Satan's doomed hordes and the Angelic hosts to battling bees:

> *Thick swarm'd, both on the ground and in the air,*
> *Brusht with the hiss of russling wings. As bees*
> *In spring-time...*
>   *... expatiate and confer*
> *Their state affairs.* [40]

ACT THREE: THE BEEKEEPER'S PART

An experienced beekeeper will provide a rest point convenient for the egg-laden queen and accessible for their capture. A water spray will induce them to settle and a hive will be waiting nearby, as only minutes may pass before scouts lead them

36 Dryden's *Virgil*, p109.

37 Butler, Ch 5 para 77.

38 Moses Rusden (1685) *A Full Discovery of Bees*, p26.

39 Gilles Augustin Bazin (1744) *The Natural History of Bees*, p178.

40 John Milton (1667) *Paradise Lost*, Book 1.1.767-775.

Fig. 8-5 : A fine eighteenth century 'embroidery in tent stitch' of beekeepers trying to capture a swarm, copied from Wenceslas Hollar's illustrations to Dryden's translation of Virgil's *Georgics*, worked by Julia, Lady Calverley, in her family home, Wallington Hall, Northumberland

© *National Trust Picture Library*

beyond reach. As previously observed, Pliny[41] recommends tearing off the monarch's wing to immobilise a swarm but Virgil seduces the insects with alluring savours of milfoil and honeysuckle[42] while Thorley[43] offers fennel and sugared ale. Ideally, they can then be brushed down from a convenient branch into a basket immediately beneath but the beekeeper must sometimes improvise!

41 Pliny, Book XI Ch 17.

42 Dryden's *Virgil*, p108.

43 Thorley, p139.

(Clockwise from top)

Fig. 8-6 : A swarm being brushed from a tree. 'Scenes of Industry', Plate 112

Fig. 8-7 : Sangfroid in capturing this resting swarm
Courtesy of Heather Leonard

Fig. 8-8 : This swarm will have to be smoked up from an awkward spot
Courtesy of Heather Leonard

Columella's instructions on catching a wild swarm are strikingly similar to American honey hunters' methods. Bees crowding around a watering hole are paint-marked and traced home. Alternatively, some may be trapped in a honey-baited reed, to be released one by one and followed. If in a hollow tree, cut the piece out and wrap it in a clean garment. If in a cave, smoke them out and beat brass instruments to scare them into settling, then capture them in a vessel as sweet smelling as Virgil's.[44] Ransome says Siberians whistle them, East Africans lead them with a bee totem; an Oregon woman writes[45] that a flashing mirror will bring them down wherever you want.

RHYMES AND TANGING

Rhymes were commonly used to charm 'God's workmen' or 'God's birds', to serve mankind and heaven. Ransome included an old German rhyme begging them to settle[46] and other rituals involved the beekeeper staking his claim by tossing Blessed Mother Earth's sand over them. Invocations of victorious women suggest links with the Mother Goddess and Amazons of the steppe.[47] Below I freely amalgamate several of the ancient rhymes.

> *SIGEWIF*
>
> *Sigewif, victorious women,*
> *Fly not to the wood,*
> *Mind my good*
> *As I do in my home.*
>
> *I beg you, mother of the flight,*
> *In God Almighty's name,*
> *Do not fly high or far,*
> *Come to my tree and claim my hive*
>
> *Prepared for all your swarm*
> *Where you can labour in His name,*
> *There set your waxen candles in His church*
> *In penitence and gratitude.*
>
> *I've strewn sweet flowers beside*
> *A ramp before your hive.*
> *Coax and drive your people in*
> *And they will serve you well.*

Aristotle knew the noisy custom of tanging with crude instruments but questions whether rattles attract or frighten bees; indeed, he questions whether they have hearing. Perhaps the practice stemmed from the legendary Curetes beating their shields to conceal Zeus's infant cries from his vengeful father, or from the clanging of cymbals at Dionysian festivals. Virgil is sure that brass and cymbals, along with sweet herbs, recall a deserting swarm;[48] Columella would use them to attract wild swarms.

44 Columella *De Re Rustica*, Book IX Ch 8.

45 *The Scottish Beekeeper*, March 1974.

46 The *Lorscher Bienensegen*, from the Vatican Library, frequently translated and quoted.

47 Annie D. Betts, 'An Old Bee Charm', in *Bee World* (1922, 4.140).

48 Dryden's *Virgil*, p108.

The Lord hissed bees for his people (Isaiah 7.18) and Saint Cyril says that was the customary Eastern way to summon them.

English instruments are more homely – kettles, keys, warming pans, hammers, shovels. Doubting the noise affects the swarm, Butler says it licenses you to pursue your claim on others' property[49] but Purchas says the law recognises your right, whatever the neighbour says. Thorley prefers to watch quietly:

Fig. 8-10 :: 'Scenes of Industry', Plate 108

When they are swarming and dancing a Levalto in the neighbouring Cloud, I never entertain them with any sort of Musick, as do the country people, which drowns the delightful and more melodious Sounds of the Bees.[50]

Anne Hughes's charming, earthy *Diary of a Farmer's Wife*[51] includes a far from docile, tuneful swarm, which appealed to me (very slightly edited) as a 'found poem'.

THE SWARM 1797

Today cum Sarah to say
The bees be swarming
So me out to find them
Setting in a gooseberrie bush,
Very wild, so Sarah did get a pot
And beat hard with a spoon
And did bang them till they did be quiet
And did hang in one lump
While we did get a skeppe to house them.

Then, taking carter's wyffe
We to the housing of them;
But la, they did go buzzing
Round her till she did dance,
Getting many stings whereon
She to the kitchen to rub some salt
To stop the stinging.

Putting the skeppe ready,
Sarah did brush them in with her hands,
Having no fear of bees,
And after covering them with sum wet wraps
To cool them down, we indoors
To divers jobs, and to dinner;

When we did laffe heartile
At carters wife who could not sit down
For the stings to her back seat,
And she did vow
Never to get near a bee again.
I fear she did feel very sore…

Mentioned in 1891[52] and again in 1908, tanging was perhaps dying out but Edwardes[53] wonders whether the noise might settle a swarm by drowning the queen's piping. This view was firmly asserted by Betsy Towner, a self-styled 'very experienced old countrywoman'. Croft[54] pragmatically suggests the noise might simply call the beekeeper home from the fields to deal with his charges.

50 Thorley, p143.

51 Mollie Weston (Ed.,1980) *Diary of a Farmer's Wife 1796-1797*, p143.

52 Letter to *British Bee Journal*, November 1891.

53 Edward Edwardes (1908) *The Lore of the Honeybee*, pp38-9.

54 In a letter to the *Daily Mail*, quoted by L.R. Croft (1990) *Curiosities of Beekeeping*, pp7-8.

HIVING AND MANAGING NEW COLONIES

Besides alluring them with scents, Butler recommends the beekeeper drink 'best beer' and also to wet his face and hands with it while hiving them. More modern details are added to age-old practice in this verse.

CAPTURING THE SWARM

Bring a skep, a sack, a hardboard square,
A queen cage and empty comb
Plus, for awkward spots, a ladder,
Saw, secateurs, feather brush and smoker.
Place the skep, shake their branch and watch.
Once tumbled in, upend it on the board
But leave a space for stragglers to get in.

In the cool late afternoon, carry it
Carefully and spill them on a cloth
Before the home you have prepared.
See how readily they march in.

Fig. 8-11 :
Transferring the errant swarm from the wheelie bin to a more seemly home
Courtesy of Heather Leonard

Casts were of disputed value, Purchas[55] and Rusden[56] both warning that too many impoverish the hive. But they can form the basis of new colonies if carefully introduced; Swammerdam claimed that in a good season, with skill, fourteen new hives could be raised from nuclei of bees and drones with their reigning queen.[57]

Another technique is to amalgamate weak colonies, overcoming their natural antagonism in a process first described by Butler. Any rival queen is removed and the queenless group will be anxious to assimilate but the others will resist. They must gradually acclimatise to each other's smells. (A modern practice is to sprinkle both groups with flour and interleave their frames – by the time they have cleaned themselves, they will all smell the same!) Then, says Butler, they will slay all subsequent Royal Issue and make a:

Happy UNION under one Prudent, Potent, Peaceful, thrice Noble SOVEREIGN.[58]

A hive that is queenless for as little as three hours will collapse socially, ceasing all work and falling prey to marauders, so a similar method is used when urgently introducing a new queen. Maeterlinck's method is still used: insert her with a few attendants in a cage plugged with candy; by the time the two parties have eaten their way through to a meeting point they are at peace with one another.[59]

An eager but well-read tyro[60] describes several efforts to modernise his parishioners' skills but found the bees had no more read his book than their keepers had! After persuading an old woman not to kill her colonies when harvesting their honey, his attempt to 'drive' them into an empty skep failed. His own swarms ended up in thorn bushes or high trees. Wishing to insert an observation pane to view a thriving nest in a cottage wall, his plan was thwarted and the colony doomed when a yokel grabbed the combs. Despite helping another cottager with his pest-ridden hives the bees died; later he heard 'they parsons are able to cast a spell on they hives'.

55 Purchas, p89.

56 Rusden, p75.

57 Swammerdam, p190.

58 Butler, Ch 5, para 68.

59 Maeterlinck, pp46-7, 66-8.

60 Herbert Brown (1923)
A Bee Melody, Ch 3.

EPILOGUE – PHILOSOPHISING

After three centuries of exciting scientific discoveries and burgeoning commercial interest, a meditative attitude returned to fashion in the early twentieth century, albeit enriched with greater factual knowledge.

Attributing the swarming moment to the 'spirit of the hive', Maeterlinck wonders why the inhabitants undertake such a risky enterprise when not overcrowded. At the apogee of their prosperity, workers' frenzied flying softens the combs and the old queen rushes about in vehement protest before they 'jet from the hive and weave an undulating, rustling veil'.[61] Laden with several days' rations plus wax and propolis, they and the egg-burdened queen rest but briefly before they must furnish their new home. Should we admire such moral abnegation to the future?

Edwardes too considers the mystery when modern hives give them all the space they need. Noting a resting swarm's extraordinary silence after deafening, joyous hubbub, he toys with the thought that depression and fear has overwhelmed them. But perhaps they have an instinct:

> *… to vary the monotony of well-doing… to break helter-skelter from the prison-bounds… go rioting one short hour of madcap frolic.*[62]

More prosaically he surmises that the behaviour harks back to a periodic need to renew dirty, worn out wild nests. Fanciful as ever, Steiner says the young queen's shining appearance disturbs the bees' twilight and destroys the power of the old queen's poison.

> *In the same way that the human soul leaves the body when it can't get enough formic acid… You can really see, by looking at the escaping swarm of bees, an image of the human soul flying away from the body.*[63]

This age-old image of soul bees morphs into the thought that the bees themselves cannot find their way into the sought-after spirit world, so they 'cuddle up together [into] a single body… in order to disappear'. Then the beekeeper's new hive can assist their re-incarnation.

A swarm in the air for so few minutes is a rare sight and an unique thrill, says Taylor,[64] filling the beekeeper with consternation at his potential loss. But how docile the bees are – thousands swirling around, a few briefly alighting but doing no harm while the mass hangs quietly and inconspicuously nearby.

Finally Longgood, another modern writer who retains a reverent sense of all creatures' mystical responsiveness to 'music unknown to human ears, as they march to drums unheard by us': dumping his first parcel of bees before their waiting hive, he watched one struggle free and raise her tail to spread the scent. Then an advance guard led them:

> *… into their new home, where the queen awaited them… like a conquering army with band and banners, taking possession of a fallen city.*[65]

61 Maeterlinck, Ch 3 *passim.*

62 Edwardes, p184.

63 Rudolf Steiner (transl. Thomas Braatz, 1998) *Bees*, p157.

64 Richard Taylor (1976) *The Joys of Beekeeping*, p39.

65 Wm F. Longgood (1985) *The Queen Must Die, and Other Affairs of Bees and Men*, p19.

9 : Bees Do It – Millennia of Mystery

Birds do it. Bees do it.
Even educated fleas do it.
Let's do it,
Let's fall in love.[1]

Given so much obscurity and intricacy, how strange that bees have become proverbial exemplars of sex education! Perhaps, simply reflecting our hang-ups, their mythic reputation for virginity explains why. They were never seen coupling, so theories abounded – for example, that they fetch their offspring from elsewhere, either spontaneously generated on flowers or produced by some other animal.

And From the Strong Sweetness

Around the ancient Mediterranean, many thought they were born of rotting carcases, a belief perhaps originating in Egyptian ox burial customs. The notion lingered from the Biblical Samson legend, memorialised on Golden Syrup tins, and, underpinned by Virgil's authority, spread through Europe into quite modern times.

But Ancient Greece offers an intricate and different explanation. The mortal shepherd Aristaeus lost his bees and all else in a flood. Angrily he asked his goddess mother to wrest a reason from wily Proteus, who says he is being punished for driving Orpheus to fathomless grief over the loss of Eurydice. So now to sue for pardon, she instructs him:

As Sirius rises in summertime
Beside four altars in the greenwood
Slaughter four peerless bulls.
When nine days have passed, send
Poppies of Lethe to Orpheus. Slay a calf
And coal-black sheep for Eurydice.

Fig. 9-1 : Aristaeus Mourning the Loss of His Bees, (1830) François Rude, Musée des Beaux Arts, Dijon
© *Courtauld Institute of Art*

1 Cole Porter

Having fulfilled all this, he delights to hear nymphs singing and bees humming in boiling clouds from the bellies of those oxen.

The Egyptian version also requires ritual precision. At each year's greening of the Nile lands, a two-year steer was sacrificed and laid on a bed of thyme, fresh cassias and broken boughs within a well-aired shed. From this ferment, Virgil unquestioningly retails:

Things of wondrous birth,
Footless at first, anon with feet and wings,
Swarm and buzz till they burst forth
Like summer showers or Parthian arrows.[2]

Such notions survived Aristotle's scepticism. The lost writer Archelaus had thought them the roaming children of a dead cow.[3] Among others' speculations involving sheep and birds, Virgil elaborates further. Fresh viscera covered with dung will produce different insects: bees from oxen; wasps and hornets from horses; beetles from asses. Another century passes and Pliny repeats all this.

Fig. 9-2 : Bees born of a dead ox, Wenceslas Hollar's engraving for Dryden's edition of Virgil's *Georgics* Book IV. Shelfmark Don. e. 187-189, Plate 30

© *The Bodleian Libraries, The University of Oxford*

In Book XV of his masterpiece *Metamorphoses*, Ovid ruminates on transience; everything, from landforms to individuals and all living matter, is constantly recycled. Bull carcases becoming bees are merely one instance of this:

The phrase 'being born' is used for beginning to be something different... while 'dying' means ceasing to be the same.

2 Derived from Greenough's version of *Georgics IV*.

3 Marcus Terentius Varro *De Re Rustico*, Book III Ch 16.

Vegetarianism, counselled Pythagoras several centuries earlier, avoids you unwittingly eating your own labourers – but overlooks the equal chance that they might be eating you.

Jewish Philo and church fathers such as Saint Augustine, even Seville's seventh century Saint Isidore, repeat these assertions, despite the flow of farming manuals. Down the ages, monasteries packed with scholars managed apiaries but their bestiaries remain unquestioning, one noting that beating dead flesh 'causes worms to form which later become bees'. Because bees abhor the smell of carrion, sixteenth century Saint Francis de Sales[4] voices some doubts. But there is still life in the old view. Shakespeare[5] remarks on a bee 'leving her comb in the dead carrion' and Ben Jonson says:

> *Beside, who doth not see, in daily practice,*
> *Art can beget bees, hornets, beetles, wasps,*
> *Out of the carcases and dung of creatures,*
> *Yea, scorpions of an herb, being rightly plac'd?*[6]

In his encyclopaedic *History of the Wonderful Things of Nature*, the seventeenth century Polish writer, Jonstonus, devotes a category to 'Things without Blood'. There he relates more fully than Virgil, perhaps more as curiosity than practical instruction, exactly how the Egyptians proceeded. They build:

> *… a house ten cubits high, and ten cubits broad, and the other sides equal thereunto; let there be but one place of entering, and four windows, on each side one; drive an ox that is fleshy and thirty months old into this place; cause many young men to stand round about him, and beat him sorely, and kill him with Clubs, breaking his very horns and bones; yet they must take heed that no blood follow, For the Bees are not bred of blood, let them not run violently upon him. Then presently stop all passages in the Ox, with clean pure napkins, dipt in pitch, as the mouth, the nostrils, the eyes, and all parts Nature hath made for Evacuation. Then laying a great deal of Thyme under, and the Ox upon it, let them come forth of the house, and presently shut the door and the windows, and daub them with Lime, that neither Ayr nor wind may enter or come forth; but the third week you must set the house wide open, and let in the light and the cold Ayr [except on a violent windward side]. The eleventh day after, when you open it, you shall find Bees hanging abundantly in clusters together; and of the Ox… nothing but his horns, his bones, his hair. They say the Kings are bred of his brain, the common Bees of the flesh.*[7]

A king might also arise from spinal marrow but brain kings are the best. Then in exquisite detail he notes the first change and transformation from inanimate grubs into living Creatures that must have air:

> *… the excrescence of their wings, yet unjoynted; and you shall see Bees in their proper colour, gathering together and flying about the king, but with small short wings, trembling for want of using to fly, and the weaknesse of their limbs.*

Evidently ideas of the development process were becoming more grounded despite the continuing mystery of the origin of bees and confusion between various flies,

4 *Introduction to the Devout Life.*

5 Henry IV Part II, IV.iii.79.

6 *The Alchemist*, II.iii.

7 Johannes Jonstonus (1657) *Bees*, Ch 11 p244-5.

perhaps particularly the yellow dung fly. Moreover, the magical notion clung on that the insects' status emanated from their birthplace.

CREDULITY, SCEPTICISM AND TENTATIVE EXPERIMENTALISM

With the rise of science, an understanding of sexual mechanics gradually displaced God's *fiat* and the old authorities. There is still delight in God's ingenuity but no obligation to accept the Bible as a manual of zoology. For instance, Jan Swammerdam found no paradox in the Samson story, reasoning that a carcase desiccating in a hot climate would soon provide a ready-made hive in its skeleton and skin. Surrendering an altogether absurd idea need not threaten faith.[8]

Nevertheless, the fancy survives Swammerdam's reproof, William Harvey's demolition job,[9] and even experiments with sealed and unsealed jars of rotting flesh that proved that only the latter produced maggots.[10] In the mid-seventeenth century 'an experiment by that great husbandman, old Mr Carew of Cornwall' is recounted in grisly detail. Replicating but simplifying the ancient method, he buries a one-year-old stirk (or steer) for ten days in spring, then exposes it under a sunny wall whereupon:

> *… it will (a great part of it) turn into Maggots… After a while, when they begin to have wings, the whole putrified carcasse would be carried to a place prepared, where the hives stand ready, to which, being perfumed with Honey and sweet herbs, the Maggots (after they have received their wings) will resort.*[11]

This 'Modern and English Experience' apparently convinced a Dr Arnold Boate to abandon his previous scepticism about Samson's story. Yet many reports of hives in skulls and helmets on old battlefields testify to bees' opportunism. Those English observations, however, suggest continuing confusion between different insect species in this country. Our climate would not assist rapid desiccation and there is no mention of combs in the carcase. In the nineteenth century, young Master Cotton gives the misconception its final twist. Being enamoured of Virgil and a born experimentalist, he found a suitable shed for a grand test:

> *I had no pity for the poor cow — no, not I — when a swarm of bees was to be the glorious result: she would surely, I thought, be happy in her death, as she would give life to so many glorious creatures. But I was not quite sure that I should be able to act the part of Guy, Earl of Warwick, the cow-killer, however much I might resemble him in spirit. I mistrusted my infant strength, and doubted much whether I could stop her nose-holes without assistance; so I straightway let the farming man into my counsels, promising him what I considered an irresistible bribe not to tell — a very small taste of my first honey; he, however, to my astonishment, did not enter into my views; my cow-killing propensities were divulged abroad, and the matter was compromised, and the cow's life spared, by the gift of a stock of bees.*[12]

8 Jan Swammerdam (1758) *The Book of Nature*, p228.

9 William Harvey *Anatomical Exertations Concerning the Generation of the Animals* (1651).

10 Francesco Redi *Experiments on the Generation of Insects*, translation of the 5th edition (1688).

11 Samuel Hartlib (1655) *The Reformed Commonwealth of Bees*. He was a seventeenth century polymath, a Polish Protestant refugee from the Thirty Years War, who devoted himself to spreading useful knowledge.

12 William Cotton (1842) *My Bee Book*, p.xlix-lii.

PROCREATION – BUT HOW?

This was still a mystery. Do monarchs, workers and drones, each of uncertain sex, generate themselves by mutual copulation? Does the monarch generate them all? Or do drones and workers mate? Robustly dismissing all such impossible notions, Aristotle's observations failed him. In his time, females were assumed to be mere feeding vessels for embryos and the left (inferior) testicle was the generator of female offspring. He sees bees as an exception, deducing that the worker is female and bears both sexes.[13] But he remains greatly puzzled, as summarised below:

ARISTOTLE'S REASONING

Absurd to suppose another animal
Would hold itself responsible
For alien offspring.
After all, why should they?

Unreasonable to suppose the worker female
And the stingless drone a male
Because nature assigns
No weapons to any woman.
Equally unreasonable to suppose
The worker male for brood is laid
When drones are absent, and no male
Habitually troubles over their young.

Evidently workers do not fertilise each other
And nor do drones. Grubs appear
Only in the presence of a monarch
Who, of necessity, must be male.
Perhaps kings unite, but what
Engenders them?

I do not know. So far will theory take us
And the limits of our observations,
But credence must always be given
To evidence from future eyes
And findings must alter theories.

13 Aristotle *Generation of Animals* Bk III Ch.X. He also observed that, when drones fly, they whirl around in the sky en masse, and then return to feast, though without knowing why. The feasting may be their normal habit but the whirling flight was presumably what was later identified as their mating flight.

14 Dryden's *Virgil*, p113.

Following Theophrastus, Virgil regards them all as genderless and virginal. Carcases forgotten, now they 'gather children from the leaves and flowers…':

But (what's more strange) their modest appetites,
Averse from Venus, fly the nuptial rites.
No lust enervates their heroic mind,
Nor wastes their strength on wanton womankind;
But in their mouths reside their genial powers:[14]

Fallon's new translation puts it wryly:

> *… you'll wonder how it ever did find favour —*
> *That is, that bees refrain from intercourse, their bodies never*
> *Weaken into the ways of love…*[15]

Despite past subtle research and his own close watch, says Pliny, nobody has ever seen their intercourse.[16] Neither can a blend of reed and olive blossoms engender the young, nor so many drones (i.e. female workers) copulate with the solitary king. They must be imperfect males born of worn-out parents. As for those called *oestrus* born of the few large cells (i.e. drones), 'noxious gadflies' that murder their rivals, they mystify him.

Saint Augustine creatively squares the biology of Noah's Ark with carcase-born sexless bees. Since only land-based creatures needed to be saved from drowning and pairs were only needed for sexual creatures:

> *… there was no need for those creatures being in the ark which are born without the*
> *union of the sexes from inanimate things, or from their corruption.*[17]

However, he adds that small creatures unnoticed in the two-by-two procession might nevertheless lurk unnumbered in the woodwork.

And so the matter rested for centuries. Finally, careful observation in the sixteenth century, the advent of microscopes and later ingenious experiments displace both myth and Aristotelian logic.

'CONCERNING BREEDING, THERE IS MUCH ADOE, AND GREAT VARIETY AMONG WRITERS'[18]

Butler's title, *The Feminine Monarchie*, makes one conclusion clear. Hawks and Amazons confute Aristotle regarding females and arms. Moreover, the Bible declares sexual coupling generates every living thing; dors (bumblebees) and chafers (beetles) do so openly. It is stupid to think workers breed workers and drones, drones, given the latter's long absence. So workers and drones must be involved in some chaste, hidden coupling.

Empirical science gains ground slowly but much ado continues. 'Like Mr Southern', Levett thinks:

> *… they doe blow as a Flye and Waspe doth in their holes or Cells… And if they are*
> *bred by copulation (as some affirme, and as may be…) … I could like of their*
> *opinions which hold the Drones to be the Males.*[19]

He knows a kingless (sic) colony fails and so does one too long without drones. That these are never slaughtered until the year's breeding is almost done shows they must have some purpose.

Purchas[20] offers his own mixture of more and less plausible theories. Some view the king as a queen who casts forth fly-blots to be nursed by drones, while others think workers contain a place for generation. He believes the huge-testicled drones

15 Peter Fallon's *Georgics IV*, lines 197-199.

16 Pliny *Natural History of Animals*, Book XI Ch 16.

17 St Augustine *City of God*, Book 15 Ch 27.

18 John Levett (1634) *The Ordering of Bees*, p61.

19 Levett p61.

20 Samuel Purchas (1657) *A Theatre of Politicall Flying Insects*, p19.

breed with the workers, whose smaller and shriller voices indicate their gender; only cows bellow louder than bulls! Workers only breed drones when weary of breeding females. He seems to be the first to notice their singular ability to know whether they are laying a male or female egg. Tradition holds that Zeus's threat to eat his children underlies their admirable chastity. Still puzzled, he concludes that 'whatsoever they do in Venus' service they act in secret'.

Topsell wastes no energy in theorising but is sure they:

> *… couple together without question as Camels do, privily and apart by themselves, which whether it proceed of modesty, or be done through the admirable instinct of Nature, I leave it to the dispute and quaint resolution of those grave Doctors, who being laden with the badges and cognizances of learning, do not stick to affirm that they can render a true reason even by their own wits, of all the causes in nature, though never so obscure, hid and difficult.* [21]

THE MICROSCOPE INTERVENES

Leeuwenhoek's instrument resolved some enigmas but he believes in preformation, every creature concealed within its progenitor since Creation. While he also retains the time-honoured belief that mothers are merely the sperm's larder, 'ovists' are concurrently asserting that the egg contains the future being.

His contemporary, Swammerdam, detailing bees' internal anatomy, establishes that the worker is a female eunuch and so useful only to labour for the community like maiden aunts – a condition that a later writer suggested might account for occasional crotchetiness![22] Drones possess seminal bladders of stupendous bulk and their parts all seem too big to penetrate the queen.[23] She has an ovary with an

Fig. 9-3 : Replica of a Leeuwenhoek microscope
© *Al Shinn*

21 Edward Topsell (1658) *The History of Serpents*, p642.

22 E. Edwardes (1920) *The Beemaster of Warrilow*, p190.

23 Jan Swammerdam (1758 ed.) *The Book of Nature*, p218.

infinite store of eggs of varying size, all tiny in the virgin. Thus he refutes many past fancies.

Some deductions have failed the test of time, one being that queen and drone cells are close neighbours so that on hatching she can promptly identify a virile partner. Swammerdam also believes the drone spends all his time enjoyably in love's labour, fertilising her eggs:

> *… in a peculiar manner, merely by odoriferous effluvia… like wanton horses neighing but not coupling like silk worms, but ejaculating like fishes.* [24]

Then finally he supposes the young queen's solitary flight expresses joy at her impregnation, rather than her search for a mate.

News of discoveries emanating from Leeuwenhoek's marvellous invention travelled slowly. After summarising millennia of conflicting opinions, King Charles II's bee-master, Rusden,[25] still clings to many traditional ideas. Rejecting Butler, he writes of a king, generals and soldiers but his tortuous reasoning leads to the familiar impasse. Others are wrong to say females receive 'a prolifick vertue from the drones' like fish, because 'the wat'ry region cannot be examples of the airy region'. Moreover, since drones are absent so long, bees would have to conceive several broods at once, like a mare delivering a succession of colts from a single mating! Secondly, a female ruler would need a husband but none has ever been spotted, or she would have to self-generate from an egg without sperm. Also, as the liquid ejaculate is unlike an insect egg, it must be a king's sperm. Lastly, bees do not copulate, so there would be no point in drones being male.

Still struggling valiantly, Rusden says conception without copulation is as impossible 'as a pair of breeches lying upon a bed got a maid with child'! But, having found sperm in a dissected monarch and watched one squirt into cells, he concludes it is male and probably can breed alone. The common bees' job is to bring in 'animable matter' to mix with sperm and create eggs. Averring that this is not pollen, he thinks its various colours provide the maternal ingredient differentiating queens, workers and drones, just as 'doggs meeting with various receptacles' produce an abundance of different dogs. The workers are female only in serving a domestic role and administering augmentation to the sperm.

Though unacquainted with Swammerdam's work, he has dissected the insects himself but somehow the drone's enormous genitals have eluded him entirely. He believes they are sexless female servants, lacking fangs for foraging and so (like capons!) only useful as nurses.

In 1702 Florinus accepts that bees reproduce by copulation but modestly conceal the act. Despite the evidence, most people deny a single queen could produce 40-50,000 offspring. Such filthy behaviour is beneath her dignity, says Jean-Baptiste Simon.[26] Giving short shrift to Rusden, Warder declares:

> *I might with as much Probability affirm that some Fly or other had cast his Seed into his Brain [and] brought forth these improvable Maggots…*[27]

Confusion continues. The notion of a male monarch proceeding from cell to cell like a 'Town Bull' scandalises him but, believing the workers are the egg-layers, he has no concerns for the queen's dignity.

24 Swammerdam, pp170, 187.

25 Moses Rusden (1685)
A Full Discovery of Bees, Ch1
passim.

26 *Le Gouvernement
Admirable*, (1740) pp xix, 4-14.

27 Joseph Warder (1722)
The True Amazons, p13-14.

PROGRESS AT LAST

The eighteenth century saw sustained efforts to resolve the mystery. Ruminating on earlier theories and humanity's fanciful attachment to bees' virtue and monarchical constitution, the meticulous scientist Réaumur[28] produced his 'Ninth Memoire, on Fecundation'. The male is full of milky stuff, the worker absolutely lacking sex parts. Confirming the queen's sole function as an egg layer, his dissections and a magnifying glass reveal invisibly small eggs in winter but a numerous inheritance before swarming. He counts over 20,000 viable eggs in one queen and over 12,000 eggs laid during March and April. As the workers build urgently, she follows close behind and backs into each void cell.

He finds it laughable to suppose modesty governs them; workers are simply artisans, licking, fanning, cleaning and feeding her. Women envy the queen's evident ability to determine drone and worker eggs for appropriate cells. There is no wanton extravagance in the number of queen eggs, whose few large cells are built with particular care, very solidly.

Following Leeuwenhoek's discovery of animalcules, maybe Swammerdam was right to think the queen is penetrated by odour – but we must assume she does what every female does. That mating has never been seen proves nothing.

He then makes determined efforts to observe sexual coupling and graphically describes the drama:[29]

I accidentally killed a laden queen, so I opened her. Having already revived her from a near drowning, I confined her in a large jar with a few of her own drones. They were indifferent and she did not need them, being full of mature eggs. So, returned from death's door, perhaps it was not surprising sex did not appeal.

Next I confined a virgin queen with an alien drone; she behaved dementedly. I was surprised to see her urgent blandishments, which he, perhaps the most imbecile of males, accepted as his due. After fifteen minutes he roused a little; she brushed his head with her arms and played with his antennae; he joined in. They curved their bodies repeatedly. She got more excited and put herself in positions which do not accord with the ideas we had wished; it would be feeble to call them other than immodest. She climbed on him and juxtaposed their tips.

After two hours I had to leave for Paris but left reliable observers there. In the whole afternoon there was no completion but he gradually became more active, put out his posterior fleshy horns and the linking arc then fell into a very long repose. She seized him with her teeth and tried to position him. But it was all useless; he was dead.

So the observers gave her a vigorous drone, who immediately did what she wanted. That evening I saw his sexual parts hanging from her and gave her some attendants. Next day I offered her another drone and put another young couple in another jar. I watched both all day; for a long time the drones were very cold, the queens very indecent in posture but perhaps not close enough to convey semen. Later, both drones died, robbed of spirit and strength, and the first queen too without a single large egg.

Undoubtedly the queen makes the advances. If the thousand males were lustful, she would have no rest or time to eat or lay.

28 René-Antoine Ferchault de Réaumur (1742) *Notes to serve for a history of insects: Ninth Memoire.*

29 Réaumur, p77-8. The passage is my translation from the French; although presented as direct quotation, it is somewhat abbreviated.

But maybe these encounters were not typical, so he speculates about future experimental designs, not realising queens had to quit the hive to mate.

Following Descartes' precept that received ideas that lack clarity and evidence should be rejected, Gilles Bazin dismisses a notion that bees bring vivifying wax home on their legs. He secures a perfect view of copulation by confining mother-bees in a glass.[30] Ignoring her own drones, a virgin queen's deferential behaviour with strangers greatly surprises him:

> … she delayed not to put out her trunk, sometimes successively to lick different parts of the male's body, at other times to offer him honey: she turned round about him, continually caressing him, either with her trunk or legs. The drone stupidly submitted to so many indearments…

After a quarter of an hour, he seemed a little animated and:

> … was determined, at length, to answer her advances, by similar ones, of the same nature. The female redoubled her vivacity, and placed herself in those positions, which agree not very well with [any] idea… of her modesty: … her indolent spouse became more active; he was animated more and more… many of those organs … appeared without him.

Over three or four hours, he found time for repeated acts of love and for repose. After too long a rest, the queen made further advances:

> … but so many marks of her repeated regard were unuseful; he was dead… She remain'd all the rest of the day fix'd to the body of her unfortunate spouse, continuing the same cares, and loading him with the same caresses… The widow of Mausolus could not better discharge her duty.

Moral Scruples

These lurid descriptions of sexual eagerness had either not reached England or fell on blocked ears. Thorley complains the queen is intolerably degraded by arguments that every drone serves her:

> As they spend not their riches in riot and drunkenness, so neither in lust and wantonness or carnal concupiscence.[31]

He worries away at the mystery. Drones have male genitals and their laziness would not be borne unless they have some function, but what? They are only present when the hive is nearly full and are killed in August, but the queen lays on and on: very strange. He rejects Purchas's proposal that she is a hermaphrodite.

> What! Does the very same Insect discharge the different Offices of King and Queen? … How can this be?[32]

30 Gilles Augustin Bazin (1744) *The Natural History of Bees*, p99ff.

31 Thorley (1774 ed.) *Melisselogia*, p30.

32 Thorley, p85.

How can Dr Warder, having seen the drone's penis of incredible form and magnitude, doubt his gender because of his inferior status? Next, he attacks Mr Bradley for wasting seven pages before conceding that the drone's organs appear formed for propagation but still denying they contribute anything! With overstrained pedantry, he then accuses Bradley of an even greater blunder, saying the king is female and therefore a queen so male is female or both sexes. This is:

> … *gross notorious Absurdity, and an absolute Impossibility! Not to be forgiven in a Professor.… And FRS.* [33]

But he agonises equally vainly to harmonise known anatomy, his convictions about gender roles and virtue, and the still mysterious process of generation. After 2000 inconclusive years, he 'will not positively determine either Way' but sets out his reasoning. God laid down coupledom but we have never seen the act. If the governing bee is male, 'then it is most evident and certain that they breed without copulation'. On the other hand, competition between so many drones would wreck the colony. Moreover, by implying the monarch is a Town Bull, Warder publicly proclaims her:

> … *a common Prostitute, a base, notorious, impudent Strumpet, the most hateful and abominable Whore, with Gallants by Hundreds. Thus has he unawares highly injured her Honour, attacked her in the most tender and affecting Part, robbed her of the most precious Possession next to her Life, ruined her Character, and stripped her of one of the brightest Jewels in all her Crown, indiscreetly exposed his queen to publick Contempt and Scorn, and shewn her no Mercy.* [34]

Should Warder be right, what becomes of her precious model of Chastity? After further worrying, he finally decides the monarch is a queen and must humbly submit to God's mysteries.

By 1796, England's scientists have evidently learned of Réaumur's experiments. Replicating them, John Keys wryly remarks that he has never seen consummation but often the queen's 'wanton gestures as would stimulate a torpedo, or any other male but a drone'. [35] He speculates that the drone may foresee and evade the suicidal act.

Precise Experiments

Meanwhile, other continental scientists continued ingenious investigations. The remarkable blind Huber [36] with his 'truly philosophic assistant' François Burnens focus on the queen's fecundity with an exhaustive series of experiments.

Swammerdam's odour theory is wrong because she remains barren when a perforated tin separates her from drones. Mr Debraw, having been tricked by sunlight glinting on debris, is wrong to believe drones fertilise the combs like frogs or fish! And he also failed to ensure his queens were virgin. Next, isolating several virgins in drone-proof cabinets, the insects disprove M. Hattorf's belief that a queen fecundates herself. Two cabinets, each containing a virgin queen, one excluding drones and the other imprisoning them, surprise them as both remain sterile. So they conclude she must fly out to succeed.

33 Thorley, p84, in a critique of Bradley's *Treaty of Husbandry.*

34 Thorley, pp92-3.

35 John Keys (1796) *The Ancient Bee-Master's Farewell*, p11.

36 François Huber (1789,1841 ed), *Observations on the Natural History of Bees*, p16ff.

They watch the mid-day hive, when the drones fly. A young queen takes flight, circling horizontally twelve to fifteen feet up before rapidly rising out of sight. They think one impregnation lasts her a lifetime but if that first brief flight is unrewarding she soon sets off again, returning after twenty-seven minutes distended by the whole male organs. Within two days, her belly is swollen and she is laying. So many apparently useless drones have heretofore embarrassed science but clearly aerial coupling carries long odds. Even so, they wonder why nature forbids mating within the hive.

Fig. 9-4 : A drone hatching
Courtesy of Paul Embden

Only worker eggs are laid for eleven months, so evidently the drone eggs are kept separately in her oviducts. Also, as a late-fertilised queen can only produce drones, it seems worker eggs must have withered while waiting. Intriguingly, those drone eggs will be laid randomly in any cell but she agitatedly refuses to lay worker eggs in drone cells when denied access to worker cells. These will be spilled in desperation, whereupon the workers eat them. But much still remains inexplicable!

The Genevan Charles Bonnet remarks to Huber that ants copulate on the wing; perhaps bees' coupling could be captured in a glass chimney. In fact Anton Janscha, eminent son of a modest Carinthian beekeeper,[37] had solved the mystery in 1771 but his important note was lost for over a century. He had detailed the mating flight, her multiple impregnations and return home with remnants of several drones on her tail.

37 After winning an art scholarship to Vienna, Janscha converted to become probably the first full-time professional lecturer in beekeeping and also Imperial Beekeeper to Empress Maria Theresa.

English science lags behind that of the continentals, though the Reverend Gilbert White, the famous naturalist of Selborne in Hampshire, identifies a regular spot where drones converge. After conjoined bumble-bees tumbled into his flower bed, Dr Edward Bevan[38] watches their proceedings in a glass for two days; there were 'no further advances on either side, rather symptoms of aversion'. Then the drone dies; the queen remains lively and flies gaily away. But the honeybee's impregnation remains obscure, though he recognises that on a warm day when drones bestir themselves she soars on her single distant journey for love.

The Victorian, Robert Huish FRS,[39] is distinctly coy about these established facts. Perhaps we have 'committed some slight offence against the rules of decency' but the subject cannot be avoided even though a publisher thinks it not exactly suitable for the female eye; but a woman has already written of it! Without any more mincing about, he dismisses Huber's views as over-rated; a single coition with a single drone casually encountered in the open air verges on the impossible. Without knowing of Janscha's findings, he is accidentally right in rejecting the single act. However, that she mates in flight is well-established and he is dogmatically wrong in declaring it:

> ... *contrary to reason... a gross and manifest absurdity to think she has stored years of future eggs. But there is a problem — she does lay when there are no drones in the hive.*

Indeed, she has a gland now known to hold 60 million sperm. But observers still puzzle as to how the mates meet. With 26,000 eyes and nearly 38,000 olfactory cavities on their antennae, Maeterlinck writes, drones from many hives will find her but only the toughest will reach her and consummate, after which their husks sink into the abyss.[40] In 2009, an article in *Bee World* magazine noted that thermal currents governed by local topography tend to create stable drone congregation areas. Ardently waiting there, they will chase off butterflies and even small birds.

Such gatherings help vulnerable virgin queens to speedy success, but how they locate the area continued to elude science until the twentieth century.[41] Tethered to helium balloons, they have now been tracked until a comet tail of drones appears. In recent years the Director of Rothamsted's Bee Research Department, C.G. Butler, established that a queen's mating pheromone, smeared on the mast of his boat, signalled so strongly that drones were still turning up nine years later.[42]

She is now known to mate several times on one flight or fly several times until filled. She gathers up to ten per cent of her capacity each time.[43] Mating impedes her agility and may cause a crash landing, only to attract more drones. Returning loaded, she strips off the male remains on her doorstep.

Detouring slightly from this convoluted trail, it was the German cleric Adam Gottlob Schirach (1724-73) who established that infant diet differentiates the 'common worm' into workers or queens, the latter receiving extra royal jelly.[44] He also thought constricted brood cells impede the workers' sex organs but we now know these are inhibited by a particular queen pheromone. Schirach was too the first modern to observe (as Bevan noted ancient Sicilians had always known) that workers occasionally lay drone eggs. This finding met with violent scepticism but was confirmed in the mid-nineteenth century by another German, Rev Dr John Dziernon,[45] should they lack a functioning queen. However, a queenless hive producing helpless drones while the workers die is quickly doomed to extinction.

38 Edward Bevan (1838) *The Honey Bee; its Natural History, Physiology and Management*, pp30-32. He knows many insects impregnate once for life; also aphid fertility is inherited. Over five generations almost 6 billion arise from one impregnation. Reproductive modes become 'curiouser and curiouser'...

39 Robert Huish (1844) *Bees: their Natural History and General Management...*, pp54-60.

40 Maurice Maeterlinck (1901) *The Life of the Bee*, p250-1.

41 R. Morse and T. Hooper (1985) *The Illustrated Encyclopaedia of Beekeeping*, p246-8.

42 R.Brown (1994) *Great Masters of Beekeeping*, p95.

43 Jürgen Tautz (2008) *The Buzz about Bees*, p117.

44 Huish, p12.

45 Morse and Hooper, p107.

ERUPTIONS OF ROMANTICISM AMID THE SCIENCE

Vanière's literary work compares the business to an Ottoman harem:

Ev'n so the Queen; she has her am'rous train;
Proud of her charms, of her attraction vain,
She boasts her Male Seraglio; inconfin'd
Her favours grants, and multiplies her kind.
When in the apt season her intrigues begin,
No law has made polygamy a sin.
…
Her paramours, an idle, recreant race
Dissolv'd in pleasure, and love's soft embrace,
In sloth inglorious loiter in the hive
…
Until their massacre at the season's end leaves
A mouldering heap beneath the frozen sky.[46]

With many complicated facts in hand but prescient about 'magnetic perfume', Maeterlinck introduces an extravagant lyricism, surely present in his evocatively named *Pélleas et Melissande* which underlay Debussy's tragic opera. The queen never mates with her own exuberant drones but:

… departs like an arrow to the zenith of the blue. She soars to a height, a luminous zone that other bees attain at no period of their life. Far away, caressing their idleness in the midst of the flowers, the males have beheld the apparition, have breathed the magnetic perfume that spreads from group to group, till every apiary near is instinct with it… She, drunk with her wings, obeying the magnificent law of the race that chooses her lover, and enacts that the strongest alone shall attain her in the solitude of the ether, she rises still… A region must be found unhaunted by birds that else might profane the mystery… Prodigious nuptials these, the most fairy-like that can be conceived, azure and tragic, raised high above life by the impetus of desire; imperishable and terrible, unique and bewildering, solitary and infinite.[47]

Bumping down to earth himself, he brutally declares her brain turns to pulp for the sake of reproductive organs, while workers' genitals atrophy to benefit their intelligence. They construct queen cells according to the reigning queen's state of fertility and ignore a newly fertilised queen until she lays her first egg.

Edwardes, however, reignites the romantic flame with his account of the virgin princess, after several hesitant forays, soaring up pursued by:

… a great drone — one of the roistering crowd that fills the bee-garden with its hoarse noontide music… At sight of him she wheels, and darts into the sunshine at lightning speed [until there is] a whole bevy of them, streaming like a little grey cloud behind her… straight up into the furthermost skies…[48]

46 Jacques Vanière (1799) *The Bees*, transl. Arthur Murphy, p22.

47 Maurice Maeterlinck (1901) *The Life of the Bee*, p250-1.

48 Edwardes, pp118-9.

10 : A SETTLED SURE ABODE

GOD-GIVEN ARTIFICE

Of course, honeybees can manage without our hospitality, building wild colonies in hollow trees and cliffs. However, that architecture of parallel, parabolic combs has also always astonished and delighted us.

Daedalus, supreme craftsman of ancient myth, made a gold honeycomb for Aphrodite's temple;[1] Saints Ambrose and Basil admired such God-given artifice. Charles Butler considers the back-to-back construction 'a woonder' and Samuel Purchas[2] asks what Euclid, anxious to demonstrate abstract lines, could do that would compare with them?

A new colony must urgently build coffers for their stores and brood. To study this hidden process, Butler took a handful of bees from his hive and watched fangs and forefeet fashioning wax but often dropping it in their haste. He thinks wax comes from flowers and also fancies the hexagons match their six feet! The lids of drone cells are like old wives' caps but the queen's palace makes a perfect figure. Placed usually on the comb edges, it is:

> *... fitting for Princes to bee neer their chief citties; yet doo they not loov to bee pestered in the midst of them.* [3]

Thorley praises the insects' most surprising and exquisite workmanship[4] and the diligence that nearly fills a new hive within a week. Believing they build with their teeth and seemingly unaware of Butler's report, meticulous Swammerdam[5] thinks he could fathom *how* in six months![6]

Meanwhile, other workers sweep the floor, seal crevices and varnish walls with propolis but already the queen is dropping her eggs before cells are built.

'FAR FROM THE COWS' AND GOATS' INSULTING CREW'

Once people interfere, their first responsibility is to provide a suitable habitat. As so often, Virgil is both practical and lyrical in setting his criteria for a bee garden. It must be protected from many dangers; for instance, the east wind god Eurus, predatory lizards and the swallow Procne. In Dryden's version, the apiary should be safe from trampling hooves and:

> *...under covert of the wind*
> *(For winds, when homeward they return, will drive*
> *The loaded carriers from their evening hive).* [7]

1 John Donald and Michael Ayrton, goldsmith and artist/ sculptor respectively, cast a modern one, in a highly skilled but not magical use of the 'lost wax' method, no doubt known to ancient craftsmen.

2 Samuel Purchas (1637) *A Theatre of Politicall Flying Insects*, p71.

3 Charles Butler (1609) *The Feminine Monarchie*, Ch 6 para 11.

4 John Thorley (1744) *Melisselogia*, p129.

5 Jan Swammerdam (1758) *The Book of Nature*, p166.

6 I describe the method in Ch 7 under the sub-heading 'Builders'.

7 Dryden's *Virgil*, p106.

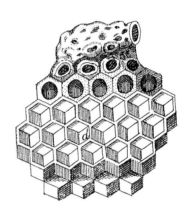

(Clockwise from top)

Fig. 10-1 : Wild comb in an English shed
Courtesy of Paul Embden

Fig. 10-2 : Detail, Swammerdam's drawing of the structure of a comb. Shelfmark Douce S subt. 48, Tab XX III
© *The Bodleian Libraries, the University of Oxford*

Fig. 10-3 : Bee chains formed when building comb or settling in a swarm. Reproduced from Gilles Bazin (1744) *The Natural History of Bees*. Shelfmark Godw. 8° 426, Plate 1
© *The Bodleian Libraries, The University of Oxford*

Fig. 10-4 : Bee garden from Virgil's *Opera*, Strasbourg 1552. Shelfmark Douce V Subt 21, fol. CIIII Verso

© *The Bodleian Libraries, The University of Oxford*

It should be full of bee-friendly flowers and watered by clear springs. The bees need stepping stones to sun themselves and drink safely from moss-green pools. Great oleasters or a palm should shade their porches. Upsetting echoes from hollow rocks should be avoided. On no account, he says, should there be yews or the smell of mire or roasting crabs!

The Roman agricultural manuals more prosaically repeat Virgil, adding a few details. Varro[8] stipulates placing hives near the villa, below their foraging grounds to ease their burdened homeward flight, and walled to withstand human robbers. Forage crops should be sown if necessary – roses, thyme, balm, poppy, pulses and rushes. Pliny[9] specifies east-facing hives to avoid both the chill of north winds and premature seduction by a southerly sun. Columella[10] proposes holes in the thief-proof wall to ease their laden passage.

Palladius repeats many of these ideas, again in Mark Liddell's splendid, muscular poetry:

8 Varro *De Re Rustica*, Book II Ch16.

9 Pliny *Natural History of Animals*, Book XXI Ch 47.

10 Columella *De Re Rustica*, Book IX Ch 5.

> *The be yerd be not fer, but fair aside,*
> *Gladsome, secrete and hoote, and fro the wynde,*
> *Square and so bigge, into hit that no theef stride,*
> *The flouris in colouris of their kynde,*
> *In busshis, treen, and herbis they may finde —*
> *Herbe origane, and tyme, and violette...*

A sobur brook amydde ellis a welle
With pullis faire; and bowis ore hit trayn
So low and rare, on hem that bees may dwell
And drynke ynough; but fer away propelle
Horrende odour of kichen, bath, gutteris...[11]

In Ethiopia, Joson dos Santos saw holes drilled in clay balls for grubs of various colours but all were said to develop into bees, *zangaons* – 'A thing, saith the author, which much amazed me.'[12] At the same time Mouffet was surprised to hear that, with no mutual disturbance, 'in the country of Abyssines under Prester John, the Bees live in the tradesmen's houses'.[13]

Cephalonia's hot, dry slopes or Provençal lavender fields are as different from Virgil's ideal as the cool green of an English garden but all satisfy the insects' and their keepers' needs. Commonest in harsher parts of Britain, many ancient walls contain bee boles. Typical are those in the walled garden of Tudor period Packwood House (see Fig. 10-6 overleaf). Like Pliny, Butler advises shading hives from the rising sun to prevent bees rousing into chilly air. However, he favours western light to help late homecomers.

He too counsels against foetid smells but says they like urine, perhaps as a physic. Grass should be mown and bushes handy for swarms. Set far enough apart to avoid bees mistaking others' homes for theirs, hives should ideally be in seven rows of nine – but 'this climactericall number' may be inauspicious, being:

... counted of some, and those no small fooles too, a parles and ominous time: more
dangerous for death, than all the other yeares.[14]

Fig. 10-5 : A Cephalonian apiary

Courtesy of Paul Embden

11 Liddell's Palladius
De Re Rustica, p6.

12 Quoted by Samuel
Purchas (1657) *A Theatre of
Politicall Flying Insects*, p168.

13 Thomas Mouffet (1658)
The Theater of Insects, p901.

14 Butler, Ch 2, para 18.

Fig. 10-6 :
Bee boles at
Packwood House,
Warwickshire
Photo by the author

Fig. 10-7 :
Nineteenth
century bee
shelter, Hartpury,
Gloucestershire
Photo by the author

The later seventeenth century saw a surge of interest in apiary and hive design. Purchas cannily faces his hives north to discourage winter ventures, because unseasonal arousal means the keeper must do more feeding. Like several contemporaries, he arranges them like tenement blocks on benches under penthouses, unafraid of confusion or quarrels. Indeed, he claims close neighbours will support and build up a weaker hive. Sometimes the whole complex may be enclosed in a shed, thus providing shelter both for bees and their keepers.[15] A late example in Gloucestershire, built in the mid-nineteenth century and unique in its ambitious ornamentation, languished unused in Nailsworth's old police station until finally coming to rest in Hartpury churchyard.

15 Purchas, p63ff.

Buckfast Abbey's Brother Adam designed his twentieth century Dartmoor apiary carefully. He is insistent that orientation does not matter, though a sunny site free of draughts and dripping trees is desirable, and one where flight paths do not cross human tracks. He distrusts bees' homing instincts and uncommonly arranges blocks of four hives facing all ways to minimise their inefficient, disease-prone and sometimes murderous tendency to 'drift' into other hives.[16]

Migrating hives to seasonal forage is a timeless practice but modern bee farming in the USA has brought new management patterns, with billions transported over huge distances to pollinate fruit farms and to maximise forage for industrial honey production.

Hives — from Clay to Cloomed and Hackled Straw or Simple Boards

John Gedde, who obtained a royal patent for his own innovations, fancifully credited Cretan King Melissus with inventing the hive. In fact, an Egyptian bas-relief of a figure working with clay hives is around 4500 years old, the earliest known depiction. Though Aristotle knew of wicker baskets with suspended combs, such horizontal, cylindrical hives were also the norm in Greece and are still found in both countries.

Fig. 10-8 : Twentieth century Egyptian clay hives
Courtesy of IBRA

16 *Beekeeping at Buckfast*
(1987), p22.

Virgil's hives were of woven osier or stitched bark with narrow doors that maintained the honey at an equable temperature. Thatched with leaves and mud-skimmed, all crevices were sealed with the same care as bees showed with propolis, a glue:

> *.... that binds more fast*
> *Than bird-lime or the pitch from Ida's pines...*[17]

Bark hives were Pliny's favourite but giant fennel or osier would do.[18] He reasons that cow-dung is the best daub because of its kinship with the inhabitants, presumably stemming from the old belief in their ox carcase womb. He speaks of Roman observation hives of transparent stone or 'lantern horn'.[19] A practical man, he recommends movable backs to facilitate inspection and further, to shrink them when honey ceases to flow, a principle that curiously disappeared for many centuries.

Varro[20] is precise – three tiers of plastered hives in a bee-house on a three-foot bank. His further words suggest a critical inventory:

> *Earthenware roasts bees in summer*
> *And freezes them in winter.*
> *Dung hives burn too easily.*
>
> *Celsus' bricks are too heavy*
> *For vital moves between*
> *Seasonal foraging grounds.*
>
> *Cork-bark hives are sound,*
> *Woven withies, fennel stalks*
> *Or hollow logs or boards.*

17 Greenough's *Georgics* Book IV, lines 44-5. Interesting to note the word 'propolis' is Greek for 'the city'.

18 Pliny, Book XXI Ch 47.

19 Pliny, Book XI Ch 16.

20 Varro, Book II Ch 16.

Several entrances confuse the eager, crafty lizard but each should be narrow to prevent pillage by either poisonous geckos or 'the foul race of beetles and butterflies and the cockroaches that shun the daylight…'.

Columella[21] uses board hives. Apiary design now seems well-established; succeeding centuries repeat many familiar things. Here are more fragments of Mark Liddell's Palladius for the sheer pleasure of the words. He rejects pottery but it 'is not angry hoot ner coold unkynde' to make the hives of 'thynner rynde', of 'saly twiggies' or 'boordis', all placed:

> *… thre foot hie on stulpis must ther be*
> *A floor for hem; wel whited thou hit see,*
> *So maad that lisardis may not ascende,*
> *Nor wickid worm, their castels forto offende…*
>
> *Her entre turne it faire upon the south,*
> *No larger than a be may trede in undir,*
> *Wikettis too or thre let make hem couth,*
> *That yf a wickid worme oon holis mouth*
> *Bisege or stoppe, another opun be,*
> *And from the wickid worm thus saf they be.*[22]

In northern Europe, around the turn of the first millennium forest beekeepers brought log hives to their own backyards. In Nepal, honey hunting, domestic log hives and modern box hives continue in tandem.

Crane records the oldest British hive, a twelfth century straw-coiled skep stitched with bramble discovered in York. No doubt there were many more in mediaeval Britain, as depicted in the margins of illuminated manuscripts.

21 Columella, Book IX Ch 6.

22 Liddell's Palladius, p61.

(Left) **Fig. 10-9** : Log hives on a contemporary Nepalese house
Courtesy of Naomi Saville

(Right) **Fig. 10-10** : Bees entering skeps. From a mediaeval Bestiary, c.1200
© *The Bodleian Libraries, The University of Oxford*

Fig. 10-11 : The last known example of wicker hives in England, c. 1890. Photo by Alfred Watkins
© *Hereford County Libraries*

Thereafter, skeps became commonplace in cottage gardens, right through to the early twentieth century. Butler preferred woven straw to cloomed[23] wicker, the whole to be topped with a steep conical hackle of straw. Otherwise any old bread crock or cream pan would serve as a cap. Wicker skeps were last recorded decaying near a Herefordshire cottage late in the nineteenth century. But there is at least one keen apiarist still keeping traditions alive by making his own straw skeps.

Fig. 10-12 : A traditionalist in his skep workshop

Courtesy of James Lamsden and photographer Rufus Reade

23 Plastered with mud or dung.

CASTLES AND PALACES GALORE

A charming detour takes us to Slovenia, to note a tradition peaking in the nineteenth century of delightfully decorated hive boards.[24] More prosaically, in seventeenth and eighteenth century Britain, smaller hives became the norm and interest in hive design blossomed, partly for easier management but also for scientific research. Some were amazingly fanciful rather than practical.

Among much else on 'the manner of Right-ordering of the BEE', Samuel Purchas[25] recommends a globe-shaped hive, which should not be too big or the bees will become lazy and dispirited. Dr John Wilkins, Warden of Wadham College, Oxford, a founder member of the Royal Society in 1660, and later Bishop of Chester, was a keen beekeeper who seemingly navigated a subtle path through political turmoil. Oliver Cromwell's sister was his wife but a taste for good living and science subsequently endeared him to King Charles. His contemporary, John Evelyn, the notable high society diarist and friend of intellectuals who also managed a prudent neutrality, records visiting him in July 1654:

> *We all dined at that most obliging & universally Curious Dr Wilkins's, at Waddum, who was the first who shew'd me the* **Transparent Apiaries***, which he had built like* **Castles and Palaces** *and so ordered them one upon another, as to take the honey without destroying the bees. These were adorn'd with variety of* **Dials, little Statues, Vanes,** *&c: very ornamental, & he was so abundantly civill, as finding me pleas'd with them, to present me one of these* **Hives***, which he had empty, & which I afterwards had in my Garden at* **Says-Court** *many Yeares after; & which his Majestie[26] came on purpose to see & contemplate with much satisfaction.*

Fig. 10-13 :
The Pilgrims' Madonna

Fig. 10-14 :
Hunter's funeral,
date 1876

Both courtesy of Museum of Apiculture, Radovljica, Slovenia

24 R. Jones (2013) *The Eva Crane Historical Collection: The Folk Art of Slovenian Hive Fronts.*

25 Purchas, p58.

26 Charles II on 30th April 1663. Pepys also saw this hive on 5th May 1665.

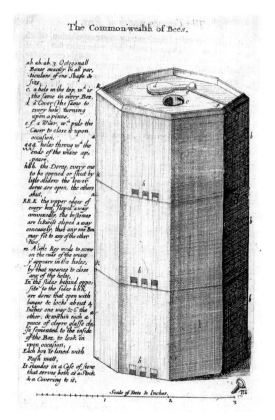

Fig. 10-15 : John Evelyn's drawing of Dr Wilkins's hive at Wadham College, Oxford, from *Elysium Britannicum*

© *The British Library Board, Add78342, image 7764*

Fig. 10-16 : Christopher Wren's beehive design, 1654. From Hartlib (1655) Shelfmark 4° H 4 Art BS, p. 52

© *The Bodleian Libraries, The University of Oxford*

The Fellows of Wadham appear to have been particularly interested in the subject, because the following year a drawing of a three-storey octagon appeared in Hartlib's papers, accompanied by a diffident letter addressed to a nameless Honoured Sir, from:

> *… a much accomplish'd and very ingenious Gentleman, Fellow of All Soules Colledge in Oxford, Mr Christ. Wren.*[27]

Compared with its contemporaries, this hive's austere appearance closely resembles one designed by William Mew but disappointingly shows none of the grandeur of St Paul's Cathedral! Should we consider Wren, a Wadham alumnus, a plagiarist? Certainly he was capitalising on his opportunities, ambition driving him 'to find any way to shew myself a Servant to a Person so eminent among the Ingeniosi as yourself'. To meet bees' habit of working from the top down, it adopted a Mr Greatrix's principle (which underlies the modern system) of inserting a second box beneath when the first is filled, and so on.

However, Hartlib included another design, even taller than Wren's – a skyscraper among hives – and so even less compatible with Greatrix's precepts. This was:

27 Samuel Hartlib (1655)
The Reformed Commonwealth of Bees.

Left for a Farewell to his Native Country, by that zealous publick-hearted and Learned Gentleman Thomas Brown, Dr in Divinity, and of the Civil Law.

Brown complains that beekeepers have forgotten ancient profitable skills, now killing their stocks to harvest honey instead of caring for them through the winter. Therefore, he sets out yet another and very detailed 'Discourse for the right making of Bee-Hives'. His New Beehive is round, 'more perfect', but flat at top and bottom with a hair's-breadth fit. It could accommodate nests eight feet long, such as wild bees make. They work themselves down the storeys and the top can be harvested without destruction. In this sweet dwelling they will prosper and in all probability:

> *... by a god's blessing, and your own moderate care, you shall have multitudes of Bees, and consequently abundance of Honey.*

Further details include comb-supporting hoops, such as gardeners use 'among their Gilliefloures', a striking forerunner of the wire reinforcements of modern wax foundation sheets.

Among this ferment of ideas about ideal hives, Hartlib reports another claim from the 'esteemed Mr Carew of Cornwall' that his colonies prospered in hogsheads or even bigger wine casks, first filling the upper part of the cask, and then working downwards. Thus:

> *... the Gentlemans usuall outcome was (through a door in the upper part of the cask) to take out what Honey he wanted, without any disturbance to the Bees, whose work and abode was then in the lower part of the cask.*

The sceptical Dr Boate, having already reluctantly accepted Carew's claim about bees breeding in carcases, was yet again surprised to think of them working in such vast spaces. He recalls the Ancients' caveat, not to make hives too large 'lest Bees should be discouraged and despair' of filling them.

In 1675 John Gedde patented yet another design. Published in a book (sold at *The King's Arms in the Poultrey* and other public houses), his title page claims his 'New Discovery':

> *... frees the owner from the great charge and trouble that attends the swarming of bees, and delivers the bees from the evil reward of ruine, for the benefit they brought their masters, advantaging the owners many fold, above what ever any method heretofore practised doth.*[28]

Clearly thinking of skeps, he believed bees mostly die within three years. He also thinks a hive full of honey induces idleness, luxury and extravagance. Moreover, an unseasonable increase in numbers must be smothered and drowned in 'the usual but unkind requital' for their devotion.

His windowed, three-tiered octagon, surely owing much to designs already mentioned, has frames for the combs. With a questionable grasp of geometry, he argues his octagon suits the insects' hexagonal cells! More tangible advantages are the space for colonies to expand rather than swarm, and a top storey that can be shuttered off to allow the harvesting of unbroken combs, untainted by smoke or

28 John Gedde (1677 ed.)
A New Discovery of an Excellent Method of Bee-houses and Colonies.

water. Interior passages for 'these pretty artists' crudely presage Langstroth's precise 'bee spaces'. He assures his customers that these industrious creatures:

> ... will requite him with double yea treble interest... Surely there is scarce any so churlish that he would not buy [them such] a house...[29]

Patronised by King Charles II, his hives were installed in several royal grounds and the design spread widely in Britain and to the continent. The Reverend Samuel Mew eulogised thus:

> ... you have built yourself a monument as well as them an house; there is not a bee which is not your debtor for her ease and life...[30]

The greatest defect of humble skeps had been the need to destroy the colony to remove the combs. It was during this ferment of ambitious box designs that the first suspended frames for skeps appeared, simplifying the cottagers' task. Despite his own proprietary design, Purchas remains content to use them; well cared for, willow-thonged straw could last twenty years and simply drying a musty hive over a fire overcomes fastidious bees' distaste. Nevertheless, he also retails a Mr Southern's demanding hygiene prescription, here presented as a 'found poem'.

> A little pease or barley be placed inside
> And a pig set to eat it while you turn the hive
> Slowly in your hands, then gently
> Wipe away his froth. Or else
> Mix fennel, sweet herbs and honey–beer
> And the bees will undoubtedly tarry.[31]

The Thorleys, father and son, published several editions of their manual *Melisselogia*. Wishing to learn from the highly reputed Dr Warder, Thorley visited his exhibition ground but felt misled by the low quality of his colonies and the poor layout, with hives:

> ... not at all agreeable, being painted with Lions and other Creatures, which I looked upon as foreign to their Improvement.[32]

He came away disgruntled but did admit the mead was very good! In later editions, the son was vigorously promoting his own bee equipment business. Small ads in the London press of 1775, for example the *Daily Advertiser*, announce:

> ... several curious Glass Bee–hives, completely filled with the finest Honey. A fine Glass Colony, full of Bees at Work, to be sold, or seen by any Customer; where may be had a new invented Glass Box Bee–hive, far exceeding any other hitherto made, both for Pleasure and Profit.

The younger Purchas's hive design is hardly new. He says the octagonal shape matches the bees' winter balling and yellow deal absorbs their damp breath. Each hive mouth should be painted a distinct colour to orientate the inhabitants. The

29 Gedde, p48.

30 Gedde, p52.

31 Purchas, p60. First recommended by Butler (1609), Ch 3, 13.

32 Thorley, p182-3.

whole should not be too heavy for the handler nor too slow to fill – because the best honey is harvested within six weeks of gathering. These hives could be bought or the pattern obtained from the *Golden Lock and Key* in Chipping Norton.

He strongly recommends Mr Jeddie's[33] terrace of hives, arboured and shelved to keep the bees safer from adversities and thus more productive of better honey and altogether easier for the beekeeper.

During the eighteenth century, bee science advanced systematically[34] and practical beekeepers continued experimenting. Condemning the 'Babelonian height' of Thorley's hives, Thomas Nutt's radical idea was to arrange the boxes laterally, even though their compartments work in much the same way as those in vertical hives.

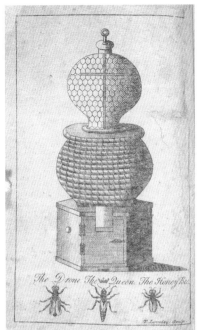

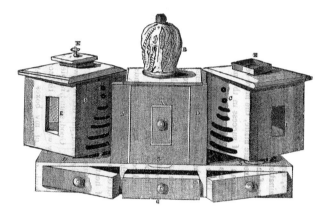

He was an unlettered but experienced Fenland bee man who found at least one noble patron, the Marquess of Blandford, who chose Nutt's hives for his palatial Delabere Park apiary. But Nutt had a large chip on his shoulder. Writing with his rector's support, he became embroiled in dispute with the 'learned Dr Bevan' and the 'big and dictatorial' *Quarterly* about the originality of his design. The rector cried shame on them and on a crude earlier structure promoted by someone named White.

> *As reasonably might they maintain that, because a boiling tea-kettle throws out steam, the tinker, or whoever he was that made the first tea-kettle, was the inventor of Watt's steam engine.* [35]

Arguments continued to rage. Bevan sardonically writes of a Huish design:

> *The principle appears to be very good, but I doubt whether it will come into general use; for as bees are not very tractable creatures, they are not likely to construct their combs in straight lines... [E]ven under tolerably favourable circumstances [harvesting] would require considerable nicety, and no small portion of courage; in some cases the difficulty would be completely insurmountable.* [36]

33 Thorley, pp179-81. There are several spellings of this one bee master's name – Geddes, Gedde, Jeddie.

34 As discussed in Chs 9 and 17.

35 Thomas Nutt (1845 ed.) *Humanity to Honeybees*, p.xxxi.

36 Samuel Bagster (1834) *The Management of Bees, with a Description of the Ladies' Safety Hive*, quoting Dr Bevan.

Since John Keys counselled women not to meddle unless clad like their menfolk in oiled linen,[37] so the solicitous Samuel Bagster offers perhaps the last fascinating quirk in this gallimaufry of hive designs. Though he approves of the Huish and White designs, they are unsuitable for the fairer sex, who ascribe legerdemain or necromancy to bee men! In fact, all they need is common sense and some knowledge. His Ladies' Safety Hive will reassure the nervous wife; each compartment is capable of being isolated as required so the bees are easily separated from their full combs at evening. Then, to take the honey next day, a 'negligee dress for a trip to the lawn is amply sufficient...'[38]

Fig. 10-19 :
Bagster's Ladies'
Safety Hive.
Shelfmark 34.17,
Title Page

© *The Bodleian*
Libraries, The
University of Oxford

THE VICTORY OF THE 'RATIONAL HIVE'

It seems that none of those designs endured. By the nineteenth century, square designs won out because they were economical to make. Nevertheless, small modifications continued to be made. Movable frames were developed by the German Dziernon and Ukrainian Peter Prokopovich; even though the latter devised spaces between the frames, the bees insisted on blocking the gaps with propolis.

At last, in 1851, the methodical American cleric Langstroth achieved a truly practical design after his eureka recognition of the fundamental importance of 'bee space'. His frames had built-in spacers wide enough to defeat their glue mania.[39]

Despite many improvers emphasising the long-term profitability of better hives, conservative cottagers clung to the skep, a decision supported on the grounds of cheapness by William Cobbett, the famous rural economist. The indefatigable Prince Albert set an example by installing modern hives at Windsor and finally Langstroth's hive became the basic pattern, with minor proprietary variants and improvements. That is still the case to this day.

37 John Keys (1796)
The Ancient Bee-Master's
Farewell, p21.

38 Bagster, p228.

39 R.Morse and T.Hooper
(1985) *The Illustrated*
Encyclopaedia of Beekeeping,
p231-2.

(Clockwise from top)

Fig. 10-20 : National hive
Courtesy of E.H. Thorne (Beehives) Ltd

Fig. 10-21 : Beehaus
© The Beehaus. Courtesy of Omlet Ltd

Fig. 10-22 : Barefoot Beekeeper's design
Courtesy of Peter Rettenberger and Maggie Wang

Fig. 10-23 : Interior view of a Barefoot Beekeeper's hive
Courtesy of Peter Rettenberger and Maggie Wang

More advances came when attention turned to the actual combs. The German Mehring's wooden dies provided labour-saving wax foundation sheets in 1857; a few years later Weiss's moulded mangles rolled continuous sheets. Now so much energy was freed for foraging that hive production soon trebled, says Maeterlinck,[40] and apiaries multiplied.

This long history gradually made life easier for the beekeeper and safer for the bees: and so the picture remained throughout the twentieth century.

'GREEN' INITIATIVES

At least two radically new hives have appeared in recent years in response to anxiety about bee populations, both designed for amateurs with small gardens and offering scope for two colonies. The Beehaus and Chandler's[41] homebuild trough employ lateral arrangements slightly reminiscent of Nutt's design. Both are said to ease swarm management and minimise interference with the inhabitants while Chandler's 'Barefoot Beekeeper' philosophy fosters natural behaviour patterns rather than the honey harvest.

40 M. Maeterlinck (1901)
The Life of the Bee, p13.

41 P.J. Chandler (2009)
The Barefoot Beekeeper.

11: LURKING LIZARDS AND COLLAPSING COLONIES

Beekeepers have always known how vulnerable their charges are to natural adversities but now we face pressing concerns about drastically declining populations.

PESTS AND NOXIOUS INFLUENCES

Aristotle's[1] list of predators includes wasps, titmice, swallows, bee-eaters, marsh frogs and toads; mildewed plants and mites sicken bees. He describes a condition now known as 'foul brood'. Destructive cobwebs soon beset idle combs. He declares they hate unpleasant smells and perfumes. Robber bees apparently cause little angst as, should they get past the guards, they so gorge themselves that rolling about they fail to make their getaway.[2]

Fig. 11-1 : Beekeepers driving away pests. Copperplate by Christoph Weigel (1654-1725) in W.H von Hohberg's *Georgica Curiosa*. Staatlische Museen
©*bpk, Berlin*

Virgil writes:

For lurking Lizards often lodge, by stealth,
Within the suburbs and purloin their wealth
And Lizards, shunning light, a dark retreat
Have found in combs and undermin'd the feat:

1 *The History of Animals*, Book 8 Ch 26 para 1.

2 If that were so, robbers would surely not be the hazard we know.

Or lazy drones, without their share of pain,
In winter quarters free, devour the grain;
Or Wasps infest the camp with loud alarms,
And mix in battle with unequal arms;
Or secret Moths are there in silence fed,
Or Spiders in the vaults their snary webs have spread.[3]

He fumigates infested hives with thyme and cuts away old wax to make the bees build new combs. Stepping stones to his stream save them from drowning. (Garden swimming pools pose a very modern problem, but a little vinegar around the edge deters suicidal tendencies.[4])

Columella[5] says echoes and fogs dismay them, and hornets in autumn. Pliny is morally censorious: wasps and hornets are degenerate and moths cowardly and ignoble. They eat the wax, foul the hive and turn cobwebs into a dangerous felt. Among odours bees dislike, the smell of crabs cooking is fatal! Before winter he fumigates with burned cow dung, but moths must be destroyed by lamps at the door:

... when the mallow begins to ripen, on a night of the new moon when the sky is clear.
Into the flame of these the moths fling themselves.[6]

3 Dryden's *Virgil*, p115.

4 Chris Koenig, *Oxford Times* 21.10.10.

5 Columella *De Re Rustica*, Book IX Ch 7.

6 Pliny *The Natural History*, Book XI Ch 21, para xlvii.

Fig.11-2 : Lizards emerging from an English hive
Courtesy of Paul Embden

Fumigating and rinsing hives with cold water every ten days in summer annoys the inmates but is good for them!

Such advice stood for many centuries. In sixteenth century Holinshed's *Chronicles*, Harrison is less prescriptive (or believes English bees are sturdier), saying only that dry hives on the warmest side of the house should stay free of mouse and moth. Thereafter a flurry of beekeeping textbooks adds and disputes little.

Butler[7] provides some colour: mice tear down combs for food and nests; the yippingal (woodpecker) pokes his long tongue in; in winter the artful titmouse knocks at the hive door until answered, then promptly swallows the bee and knocks again. Shocking to modern ears, he instructs boys to destroy their nests. Spider webs entangle weary homecomers but a healthy colony can outface emets.[8] Snail slime and mealy caterpillars offend them. But worst is the human thief, who is more despicable than highway robbers or sheep stealers and deserves to lose the Muses' favour!

Fig. 11-3 :
Woodpecker damage to a modern hive
Courtesy of Paul Embden

Fig. 11-4 : Bees mobbing a robber wasp
Courtesy of Paul Embden

7 Charles Butler (1609)
The Feminine Monarchie,
Ch 7, para 1ff.

8 Emet, like pismire, is an old or dialect word for an ant.

9 Samuel Purchas (1657)
A Theatre of Politicall Flying Insects, Ch 20.

Purchas considers wasps a greater menace than hornets, as they graduate from eating dead bees to raiding the hives. Pismires, too, eat the honey but swallows, toads and woodpeckers are more maligned than guilty.[9] Topsell warns against foul smells emanating from 'houses of Office' and kitchens, before waxing eloquent about the presence of immoral people. It is necessary to:

... remove from their Hives mouthes, unlucky, mischievous and deceitful people, and idle persons [or those] stayned with whoredom... or gonorrhea, or the fluxe of menstrues.[10]

Good management and box hives can prevent most dangers, says Rusden, but Thorley highlights 'their most ingrateful, unjust, cruel and merciless Owners' as the chief hazard.[11] Earwigs thieve honey but sparrows' liking for brood is no worry. Snails probably cause little harm; he queries a tale of one stung to death and then, too heavy to move, encased in glue.[12]

Unlike Purchas, the eccentric Oxford biologist Frank Buckland (who notoriously ate the most unlikely specimens) asserts that 'toads are capital hands at eating bees'. One owner found one with a stomachful.[13]

Three twentieth century writers bring new observations. Brown, perhaps fancifully, declares that bees so dislike the scent of foxes that they welcome the hunt.

When, therefore the huntsman's horn sounded, and the scarlet coats flashed amidst the dark undergrowth, and the hounds bayed as the fox broke cover, the bees flew the more rapidly, expressive of joyful mirth at a riddance of a pest...[14]

Taylor[15] finds it droll that skunks too scratch at hive doors at night to snatch whoever answers. The patient and Buddha-like toad also amuses him, with its confidence that some insect will come within range before it starves. He delights in snakes finding warm havens under his hives, and a nesting hen too with her chicks protected by legions. Lastly, countering ancient wisdom, Longgood remarks that manure water causes 'a frenzy of excitement and joy'. Clearly:

... bees do not read standard texts. At least my bees. They invariably show a preference for water that is stale, green and slimy with algae... [or] dark, festering, sinister brews that would probably paralyze or kill humans.[16]

Morse and Hooper's modern textbook dismisses many supposed bee enemies as occasional, harmless lodgers or minimally tiresome freeloaders that sometimes usefully scavenge or prey on nuisance intruders.[17] Some parasitic flies, beetles and problem mites or wax moths in the hive are a real pest but outdoor predators rarely warrant concern. At least one modern beekeeper[18] positively welcomes spiders to catch the moths and finds his bees licking winter-sheltering slugs for moisture to help them eat crystallised honey.

10 Edward Topsell (1658) *The History of Serpents*, p646.

11 John Thorley (1744) *Melisselogia*, pp163ff, 177.

12 Thorley, p187. No doubt the antiseptic propolis with which they embalm or seal up offensive litter.

13 L.R. Croft (1990) *Curiosities of Beekeeping*, quoting from Francis T. Buckland (1874) *Curiosities of Natural History*.

14 Herbert Brown (1923) *A Bee Melody*, p20.

15 Richard Taylor (1976) *The Joys of Beekeeping*, pp68, 71.

16 William Longgood (1985) *The Queen Must Die, and Other Affairs of Bees and Men*, p39.

17 Roger Morse and Ted Hooper (1985) *The Illustrated Encyclopaedia of Beekeeping*, pp291-4.

18 Personal communication from Mike Hicks, Scilly Isles.

Fig. 11-5 :

Bees investigating robber Death's Head Hawk Moth. Its 'peeping' sounds like their queen's puzzles the bees

Courtesy of Paul Embden

DISEASES

Stricken with disease, according to Virgil bees lose all brightness and show:

> *… mournful pomp;*
> *Or foot to foot about the porch they hang,*
> *Or within closed doors loiter, listless*
> *From famine, benumbed with shrivelling cold.*
> *Then is a deep note heard, a long–drawn hum…*[19]

Then, he says, it is time to burn scented galbanum (gum resin) and tempt faltering appetites with a prescription of honey, flavoured with bruised gall, dried rose-leaves, must, Psithian raisin-grape juice, bitter centaury, Cecropian thyme, and star-wort roots well seethed in fragrant wine.

Pliny[20] says gluttony causes flux and looseness in spring and late summer. If hideous and shrunken or silent and mournful (which sounds very much like spring starvation), then he uses Virgil's nostrum but with Aminean wine. Disease being so common, Columella[21] stresses that new swarms must be caught in spring. As for resurrecting dead bees, older authorities would lay them in a dry place through winter and cover them with fig-wood ashes in the sun at the spring equinox, but he prefers a dietary cure – pomegranate with Aminean wine, or raisins, sumach and wine. Or rosemary in honey water, or ox urine!

English writers describe similar symptoms. Butler notes how the insects hang on each other's heels and hum mournfully, to be cured by warmth and food.[22] Mouffet elaborates – 'loss of mirth, lumpish melancholy, vertiginous or whirling gate or motion [and] rough physical appearance'.[23] Purchas denies their fragility; decent food and a decent site will keep them healthy.[24] Recognising only seasonal flux, Rusden blames this on a lack of vegetables, to be treated with a salt and honey dose or perhaps meal and water, either remedy sounding less appetising than the Mediterranean cures![25]

Like the Ancients, Warder[26] claims chilled or sick bees can be warmed beside the fire and dead ones raised to life on warm ashes, whereupon, 'opening my Handkerchief, they have all flown Home, every one to his own Hive, as readily as if they had never been dead'. Resurrection from April famine takes longer; having placed the bees on plates on a sunny wall, when he sees legs begin to waggle, he spreads honey on their combs, returns them to their hive and after a week they are fit to work again.

Clearly winter starvation was insufficiently understood, though Warder stresses the need for proper stores in September and feeding in April. He also warns against waking them with dribbles of food in hard weather.

When a colony is sick, observes Brown with romantic melancholy, hundreds deliberately:

> *… leave the warmth and comfort of the hive, run with haste down the alighting*
> *board, and betake themselves to blades of grass, where, with upturned faces, dying,*
> *they gazed upon a superb sunset.*[27]

19 Greenough's *Georgics*, Book IV lines 428-52.

20 Pliny, Book XI Ch 21.

21 Columella, Book IX Ch 9.

22 Butler (1704 ed.) *The Feminine Monarchie*, p14.

23 Thomas Mouffet (1658) *The Theater of Insects*, p904.

24 Purchas, p123.

25 Rusden, p131.

26 Joseph Warder (1722) *The True Amazons*, p113.

27 Brown, p166.

Weather Hazards

Adverse weather is an eternal enemy, causing casualties in flight and the loss of whole hives in winter. Sound hives will stay dry inside and out, says Butler, but frosty sun may tempt bees out too soon. Then in a doorstep melée they tumble over each other and die of chill unless a warm hand or fireside revives them.[28] Sparkling snow attracts but dazzles, says Purchas, while winds can defeat them when laden.[29] Brown[30] speaks of them clinging numbly to flowers, suddenly trapped by an east wind on a fine May morning.

Mouffet reports how night thunder and lightning frighten them. While with Dr Penny attending the Countess of Somerset's death-bed, they heard an unknown colony of bees agitating under the next ceiling:

> ... a great noise, as it had been an alarm of war... The next day for want of sleep they flew about making a hoarse noise, trembling and not knowing what they did, [dashing against windows and stinging each other].[31]

'The Trumpet Sounds and the True Sons of Mars Prepare'[32]

Bees have their own formidable armoury. Aristotle remarks that their stings can kill a horse. In 1914, J.H. Lovell confirmed a traditional belief that bees were averse to the colour black, finding seven such chickens stung to death while their paler sisters escaped. Similarly, black horses among a group of light ones were more often stung and the black patches of a piebald cow were struck five times as often.[33]

However, it is the solidarity of a single hive and the warfare between them that most amazes and perturbs. Butler[34] reserves his greatest wrath for robber bees 'at perpetuall defiance and deadly feud', 'freebooters sending forth their stoutest yunkers' against weak colonies when nectar supplies are shrinking. If repulsed, they summon reinforcements with:

> ... such a noise and din as if drumbeat of alarm or trumpet sounded a charge or a shrill flute...

A fully clad beekeeper may quell the combat with water-spray and defeat the invaders, whereupon the residents gather at the gates, talking of the fight. Such battles signal time to take off his share of the honey and narrow their door.

Rusden too notes these perilous late season raids and, with unfortunate consequence for the brewer, has known occasions when:

> ... being sensible of their own poverty, they have been led into brew-houses by the sweet smell of the wort, where many bees have drowned...[35]

Thorley seems morally confused, deeming these robberies not driven by cruelty but by 'perfect Abhorrence of Sloth and Idleness' and 'insatiable Thrift'. He nevertheless abominates such a:

28 Butler, Ch 3 para 1, Ch 7 para 63.

29 Purchas, p121.

30 Brown, p24.

31 Mouffet, p899.

32 Thorley, p163.

33 *Entomological News* (1914) 29 (9) 407-10.

34 Butler, Ch 7 para 25ff.

35 Rusden, p123.

... martial, unpeaceable Spirit, and notorious Injustice, striving to enrich themselves, tho' at the expense and ruin of their neighbours... 10,000 pities that such excellent and useful creatures should thus plunder and destroy each other.[36]

Hordes follow a successful vanguard and battles can rage for two days until the hive is ruined. A wand of stinking madder (a plant of the *rubia* family) might stir puny residents to greater valour but no beekeeper should endanger himself by defending a weak hive. In one particularly malign August, the majority of Thorley's thirty hives were lost or damaged.

Bazin[37] reports some individual fights lasting an hour and a half, one bee dragging the other from the hive and seizing it wherever it could:

... try'd to mount upon its body; when it was fixed there, it seized the conquered Bee by the neck, and strangled it with its teeth... [until] lifeless in the dust, or ready to expire there: then she abandoned it, but continued settled near her, as if to enjoy her victory, in rubbing itself with its two hind legs, as a man rubs his hands, when he has done something, with which he is satisfied.

Interestingly, he describes quarrels designed to recover the loot but not to kill. Having often seen several bees in cowardly fashion harassing and biting a single one:

I own, at first, I had pity on the unhappy wretch...; but when I observed the Bee... had an easy method to free herself from them, I understood they had no design on its life. The combat was at an end, when the creature, thus bit and tormented, put out its trunk: for immediately one of the aggressors came to suck it, by applying to its own trunk, as did the others in their turn; so that all these Bees seemed to have no other end in their attack, than to force her to disgorge the honey, which she had refused them.

36 Thorley, p163.

37 Gilles Bazin (1744) *The Natural History of Bees*, pp77, 78.

38 *The Guardian G2*, 25th May 2010. Extracted from Gavin Pretor-Pinney (2010) *The Wave Watcher's Companion*.

39 Rudolf Steiner (1998, transl. Thomas Braatz) *Bees*, pp177-8.

40 Mark L. Winston (1998) *From Where I Sit: essays on bees, beekeeping and science*, p51.

41 Fiona Harvey, *The Guardian* report, 10th March 2011.

42 Peter Graystock *et al*, 'The Trojan Hives: pollinator pathogens, imported and distributed in bumblebee colonies', in *Journal of Applied Ecology* (online) 18th July 2013.

In 2010, an article in *The Guardian* described wild Asian bees' defensive tactics. When hornets threaten their cliff-hanging combs, they perform a 'Mexican wave'. Flicking their tails in the air to reveal darker undersides, yellow and black ripples shimmer in an expanding spiral pattern that appears very large and menacing.[38]

DISAPPEARING BEES: THREAT OF A CROP CATASTROPHE?

Rudolf Steiner's feet were perhaps on the ground when he forecast disaster from modern queen breeding.[39] In England, Brother Adam spent a lifetime seeking a breed that ideally balanced health and hardiness against easy management and productivity. In North America, there are fears about feral Africanised bees mating with queens bred for docility, possibly resulting in hives 'much like inner-city gang[s], dominated by a few aggressive leaders'.[40] In 2011, the UN Environmental Programme noted how some well-intentioned interventions – breeding programmes and massing bees in huge hives – can exacerbate some problems. Accelerating global trade also spreads pests and diseases.[41] Bumblebee colonies now imported for crop pollination show a high percentage of parasites, posing a significant risk to resident pollinators.[42]

Undoubtedly, falling bee populations are the biggest issue as I write in 2014, concerning apian scientists, beekeepers and farmers. Other pollinators are also suffering badly but honeybees pollinate at least thirty percent of crops. Having relatively few immune or detox genes,[43] they are prey to endemic varroa mites but important questions arise as to why their resistance has recently reduced.

Charles Darwin offered a fascinating ecological insight. Honeybees are anatomically unsuited to foraging red clover, so that crop depends on the prevalence of bumblebees; their ability to thrive depends on a scarcity of field mice eating their combs, which in turn depends on the control exerted by the local cat population. Thus, ultimately, red clover depends on cats![44]

North American commentators refer to 'colony collapse disorder' (CCD), a label rejected in Britain where losses, though very serious, have been less precipitate and catastrophic. Unlike other ailments, it happens suddenly and leaves hives virtually deserted with no sign of sick or dead bees in the vicinity.

Unlike timeless migratory beekeeping, in the United States billions are trucked many thousands of miles to pollinate a succession of crops. Does this industrial scale migration overstress them? The practice is long-standing, so owners look elsewhere for explanations. Complex interactions between several adversities – antibiotic resistance among their pests, viruses, environmental chemicals and an impoverished diet from less varied forage – are suspected, on both sides of the Atlantic.

STRONG MEN CRYING

Men with names like Matt and Jim
And a girl who's taken dead Dad's place,
With trucks bigger than Amtrak cars,
Prepare to drive their dormant hives for days
Across the States to California.

Around kitchen tables tensions rise —
Anxious wives add up the books,
Hide tearful eyes behind their hands,
Whisper to unwilling ears, It can't go on.
Last year 800 colonies died. Unless
All you take survive this spring
We're busted. *Tartan-shirted men*
With brawny arms reply, I love
My girls; this is what I do, it's me.

A red and yellow sunset fades
To pink and turquoise. Peace.
Trucks congregate in farmers' yards,
Their loads are spread under the still
Wintering almond trees. I mark
Each hive – who's dead, who's alive.
A seethe of black around the combs
In some; a wretched dust falling

43 US Dept of Ag Bee Research Lab and ANU, Canberra Research School of Biol Sciences (2006).

44 Longgood, p51.

From too many, exhausted remnants
Struggling to push their sisters' corpses
A decent distance from their gate.

The sun is bright. Blossom bursts,
Painting the world and latticing the sky
With white. Mouths probe among pink stamens,
Legs load with gold, wings whirr
And pollen spreads. They forage
And gather nectar as though
Nothing changed. **Too few.**

Shoulders droop, men bury their heads
In the steering wheels of those monster trucks
And refuse to tell their wives back home.
Urgent parcels of fresh Australian bees
Are spilled into sepulchres. **Will they survive,**
Revive a colony? *Or is their doom*
Already sealed by this plague?

Table talk in the motel is all about
Bills that won't add up. New stock costs
More than they'll be paid, a gamble
On a better year next year. If not, the end.
But who would buy? **I love listening**
To their hum. I just want to carry on;
This is what I do, it's me.

THE FIGHT BACK?

This dearth of pollinators has triggered widening research, ranging from the effect of chemical cocktails on their navigation or dance interpretation to breeding disease resistance and diversifying forage. But is it enough? The British government reported in 2008 that honeybees contribute £165 million annually to our agricultural economy and billions worldwide; although much higher values are mentioned occasionally, I have not found an authoritative update. Can money best measure the ecosystem?

In 2000, Dr Denis Anderson[45] of Canberra tracked back to a single Korean mite, *varroa destructor*, which infected a European bee in mid-century and then quickly spread through that economically dominant and universal species. His hope is to interrupt the mite's reproductive trigger before it takes hold in Australia, the only country still able to resupply clean bees worldwide. UK surveys find fluctuating numbers, variously attributed to changing weather patterns inhibiting foraging while unusual cold kills off disease agents. The British Beekeepers Association focuses on improving keepers' skills and highlighting habitat destruction that reduces forage.[46] Others hope to breed bees resistant to endemic varroa mite and other serious illnesses.[47]

45 Of the Federal Government's Commonwealth Scientific and Industrial Research Organisation, Canberra.

46 Various media reports, October 2010.

47 National Bee Unit, York; BBKA; honeybee1.org

Populations have reportedly improved where certain pesticide sprays were banned, nerve agents accused of causing a condition the French called 'mad bee disease'.[48] Despite some hesitancy in the UK, vigorous campaigning recently led Europe to impose a two-year moratorium on neonicotinoids.[49] Scientists doubt this trial is long enough to clear pesticide residues from the environment and enable reliable findings to emerge. Systemic substitutes may pose their own risks to foragers or ground-dwellers and the BBKA fear a reversion to more destructive agents. Meanwhile, many farmers fear greater losses to pests. The government is urged to 'get on' with research into alternative pest controls[50] and has promised a national pollinator strategy.[51]

Countermeasures for Everyman

Desperate bees may be taking up their own cudgels. US entomologists have found them sealing cells containing contaminated pollen but colonies are still declining.[52]

In recent years some British farmers, having seen urban gardens' richer forage, have started establishing flowery areas among crops and reaping better yields than sterile modern fields. Frequent news stories about the bee crisis have caused enrolments on apiarist courses to shoot up, along with bee-friendly planting and garden or rooftop hives.

48 Rigorous studies by teams at Dundee and Newcastle Universities have now established that learning centres in bees' brains are damaged by neonicotinoids, seriously compromising their capacity to identify food.

49 Announced by the European Commission on 29th April 2013.

50 BBKA news item, 29th April 2013.

51 Lord de Mauley, Environment Minister, speaking to Friends of the Earth, 28th June 2013.

52 US Department of Agriculture and the University of Pennsylvania, 5th April 2011. Jeffery.Pettis@ars.usda.gov

Fig. 11-6 : Public flower bed, 2013
Courtesy of Friends of the Earth, Eastbourne

More local authorities, supermarkets and commercial organisations are offering practical support to beleaguered bee populations. For example, leading on various green initiatives, Newcastle-on-Tyne's Bee Strategy involves replanting municipal beds with bee-friendly flowers and installing more hives.

Governments and other organisations, including the BBKA, support improvement projects in developing countries in various ways: helping people to produce and market saleable products; forming co-operatives to obtain better equipment; training them to use frame hives. Fig. 11-7 shows a simple hive that protects against invasive pests on display in the Botanical Gardens of Kathmandu University, Nepal.

Fig. 11-7 : Frame hive in Kathmandu Botanical Gardens
Courtesy of Heather Leonard

DEAD BEE ON THE WINDOWSILL

His bees have failed, so Chen Wong–Li
And all his kin climb their groves of trees,
Every mandarin and lychee,

Brushing pollen from each flower
Stamen to stigma to ensure
The only way they can, secure

Harvest and livelihood.
Dropping school and greenwood
Hopes, a bellyful of food

Is all the compass of his children's minds.
Arm's length plus a feather in the hand
Describe such inelastic bounds.

If we fail to save the bee
And lose the work they do for free,
Not only Chen Wong–Li

But all of us will not know where
To turn for food. We must play fair
With this our world or face despair.

Let us end with this optimistic bee sculpture at the Eden Project in Cornwall.

Fig. 11-8 : Bee sculpture at the Eden Project

Courtesy of Tim Smit, Eden Project

12 : HONEY

'SWEAT OF THE HEAVENS... SALIVA OF THE STARS'

Such was Pliny's description. Countless myths tell us that a diet of ambrosia made gods and men immortal – but that sublime substance was not always honey. The word could mean nectar, mead, or even honeydew, now known to be exuded by aphids but which Butler calls 'a Quintessence of all the Earth's Sweetness, condensed by the sun and falling like fine rain'.[1]

In reality, nectar is honey's main ingredient and the pollen that bees gather is a crucial protein, a particularly important constituent in royal jelly. Flower species, weather and soil type all affect the quantity and quality of both. Different pollen colours are obvious on bees' laden back legs. Their third target is resinous glue from trees and sticky buds, the propolis with which they seal and disinfect their hives.

People first thought they simply gathered honey ready-made from flowers or, Aristotle says, falling from heaven when the stars rise or the rainbow descends.[2] Seneca was the first to wonder whether the insects actually contribute something themselves but even Swammerdam is unsure; does their digestion merely modify the gift's consistency or do glandules in their trunk somehow change it?[3] Not until 1717 did the French scientist, Vaillant, demonstrate that the flowers' product is simply nectar.

We now know that bees' subtle chemistry creates a substance of marvellous complexity: seventy percent various sugars and twenty percent water, but the final ten percent contains around 200 other substances – vitamins, minerals, proteins, acids, enzymes, and volatile components including alcohols and esters. Their forage determines the exact mix and final colour.[4] An outlandish multicoloured comb once resulted from a raid on a seaside rock manufacturer![5]

Perhaps more bizarre are Herodotus's two reports[6] of man-made honey. Around the River Meander, scene of *The Iliad*'s bloody battles, Xerxes' army found craftsmen making honey out of wheat and tamarisks. Among Libyan nomadic Gyzantes, who paint themselves with vermilion and eats apes, he speaks less certainly of much honey made by bees, 'and much more yet (so it is said) by craftsmen'.

WHERE THE BEE SUCKS...

Virgil's tale of air-born honey[7] itemises his bees' ideal garden, starting with 'green cassias and far-scented thymes, And savory… and violet-beds'; later pages include saffron, thyme, roses, endives, parsley, narcissus, acanthus, myrtle, lilies, vervains, poppies, pines, lime trees and orchard fruits.

Allowing for some Mediterranean and Northern variations, bees' tastes naturally remain consistent through the ages. Levett[8] and Purchas[9] list early blossom trees and a progression of later plants. Purchas notices that the longer tongues of bumblebees mean they can forage where honeybees cannot. He quarrels with others about their palate – of course the flowers of 'bitter' mustard and radish are sweet and therefore appealing. (I wonder whether bees read his warning list of noxious plants such as

1 Charles Butler (1609) *The Feminine Monarchie*, Ch 6 para 41.

2 *History of Animals*, Book 5 Ch 19 para 4.

3 Jan Swammerdam (1758) *The Book of Nature*, p161.

4 Bee Wilson (2004) *The Hive*, p146.

5 Wilson, p147.

6 Herodotus *Histories* 4.194.1 and 7.31.1 (pp335 and 456 in Penguin Classic edition).

7 *Georgics IV*, line 1; bee-friendly plants scattered throughout.

8 John Levett (1634) *The Ordering of Bees*, p51ff.

9 Samuel Purchas (1657) *A Theatre of Politicall Flying Insects*, p94-5.

yew, box, spurge, wormwood, woad and wild cucumber.)

Despite a somewhat dilettante reputation, the diarist John Evelyn had a keen interest in gardening and admired Sir Robert Clayton's grand new home with its garden 'so full of Wild Thyme, Marjoram & other sweete plants' and nearly forty hives.[10] His *Sylva* offers a woodland perspective; elms may cause dangerous dysentery but oaks do not deserve their reputation for causing bees to gorge themselves to death. Valuable early nectar sources are black cherry, willows and buckthorn. He notes Trebizond box honey's regrettable tendency to drive its consumers mad.[11] These woodland recommendations and warnings offer little to most people's modest plots but, surrounded by 'improved grassland' and grubbed-out hedges, bee-friendly private gardeners are vital and the RHS lists seasonal recommendations from early hellebores to ivy's final offering of 'bottled summer'.

The Busy Bee

Foraging is astonishingly labour-intensive; around 1500 flowers provide a single bee's pollen cargo and a colony flies the equivalent of three times round the world to fill one honey jar. Disproving speculations about their reach, Huber's marked bees traversed as directly as a cannon ball within a half-hour radius, shorter where forage was rich or competitors few.[12] Modern investigators have found that a mile radius suffices in a good area but they routinely fly up to four miles. An experiment in the Wyoming badlands proved the ultimate boundary was eight and a half miles from the hive.[13]

10 John Evelyn *Diary*, 12th October 1677.

11 See Pontic honey, p154 and Chapter 16.

12 François Huber (1789) *New Observations on the Natural History of Bees*, p183-4.

13 R. Morse and T. Hooper (1985) *Encyclopaedia of Beekeeping*, p134.

Fig. 12-1 : Bee-friendly English countryside

Courtesy of Paul Embden

Unlike butterflies' random flitting or the single species diet of various solitary bees, their 'flower fidelity' makes honeybees particularly efficient pollinators. Edwardes[14] fancied that fashion or sex may influence this but modern studies have shown it serves two functions. First, they learn the markings of good flowers and what techniques win nectar from different species. For instance, they will use holes made by bumblebees at the base of a bean flower; they gradually learn to enter lucerne from the side to avoid being trapped; they know the different times of day when different nectars or pollens peak.[15] Secondly, as different contingents from a colony target different sources they achieve an overall balance in pollen nutrients.

THE ART OF HONEY TASTING

Ancient writers displayed critical palates. Ovid claims the nobility of the carcase that produced the bees determines the honey's quality![16] More persuasively, flower and season are the main measures, summer honeys being choicest. Mount Hymettus thyme excites endless praise from ancient Greeks but disappoints a nineteenth century German traveller who rues a land denuded by wars and shepherds; he most prizes Cycladean honey.[17]

Campaigning around the Caspian's evocatively named Fortunate Villages, Alexander the Great[18] finds wild combs of surpassing sweetness. The geographer Pausanias[19] gives top marks to Hymettus honeys but also meets the toxic Pontine product.[20]

Pliny[21] is a connoisseur, declaring the best product comes from countries with the most exquisite flowers, for instance Sicilian Hybla, and the cleanest from a fly-free mountain in Cyprus. Heather honey is poor stuff. Oak, linden and reeds offer the finest flavour but thyme honey is the most delicious and serviceable, with ideal viscosity, fine aroma and an interesting 'sweetness that closely borders on the sour'. He marvels over large combs from the north, particularly a German specimen eight feet long! He warns against dark red Pontine honey, which makes people sneeze, develop fevers and throw themselves to the ground. It being so dangerous, taxes there are levied not in honey but wax.

Fascinating to reflect that the Pontic region is the home of the legendary sorceress Medea who killed their children and poisoned his bride to avenge Jason's infidelity.[22] Perhaps her name, like so many other honey words, derives from the Sanskrit *madhu*.

Praised by the ancient poet Kuo P'o, China's abundant and various honeys are noted in Purchas's *tour d'horizon*, 'for they delight in keeping of bees'.[23]

'Honey isle' was a Druid name for Britain, where wild thyme is as plentiful as in the Mediterranean. Harrison[24] boasted that Elizabethan England produces the best honey. Cleaner processing breeds less choler; foreigners, he has heard, stamp on their combs, allowing 'bees and young blowings altogether into the stuff'.

Butler joins the chorus praising thyme honey but observes bees' pell-mell rush for honeydew on oak leaves, stumbling into each other at the hive door. He differentiates liquid nectar from solid ambrosia (valuable but fast-decaying pollen) and comments on liquid 'liv-honni' hardening into 'ston-honni'. Uncapped late season honey is thieved repeatedly by robber bees – 'Evil gotten goods are soon spent.' Like Pliny, his favourite honey:

14 Edward Edwardes (1908) *The Lore of the Honeybee*, pp58-9.

15 Why don't other insects also learn? Are honeybees singularly intelligent?

16 Ovid *Metamorphoses* xv. ll.365ff.

17 William Cotton (1842) *My Bee Book*. Appendices.

18 Diodorus *Historical Library* 17.75.7.

19 Pausanias (C2CE) *Description of Greece*.

20 From *Azalea pontica* and *Rhododendron ponticum*. Aconite honey too can produce vertigo and delirium; also *kalmia latifolia* in USA, which killed many Philadelphians in 1790. See Ch 16.

21 Pliny *The Natural History of Animals*, Book XI and Book XXI.46.

22 Euripides *Medea*.

23 Purchas, p140.

24 William Harrison (1587) *Elizabethan England*, p175.

… is clear, mellifluous, yellow like pale gold (but right Virgin honey is more crystalline at the first), sharp, sweet, and pleasant to the taste, of a mean consistency between thick and thin, so clammy that being taken up upon your fingers end, in falling it will not part, but hang together like a long string. [25]

A good English harvest would be up to three gallons per hive, so Pliny's report of the huge German comb amazes him,[26] as do Paulus Jovius's reports of great forsaken lakes of honey in Muscovy woods. Amused, he recounts a Russian Ambassador's story of a neighbour, '… slipt doun into a great hollow tree and sunk into honni up to his brest;… stuck fast two days…' until serendipitously rescued by a bear.

IN ANCIENT MUSCOVY[27]

We toast Dmitri newly here in Rome,
Startling the company with tales from home.
Should we believe this gruff ambassador
Who tosses empty flagons to the floor?

He praises endless forests full of bees
With wells of honey filling hollow trees,
Of rough peasants climbing with sharpened knives
To plunge their fearless arms into wild hives.

We roar with laughter and raise a full glass
At Misha's strange fortune, a dreadful pass
As he stumbled feet first and crashed chest-deep
Through the comb he had set out to reap.

Stuck fast in the honey and no-one knew,
Poor man, the story takes a darker hue —
Two days he hollered as loud as he could
But no-one could hear him so deep in the wood.

And then a brown bear, drawn by the sweet scent,
Lumbered in backward, claws first in descent.
The man in such mortal fear either way
Could hardly catch a lung of air to pray

As he **beclipt him fast about the loins**
And made an outcry *bold. The startled* **Bruin**
Heaved free of the mire while Misha clung tight.
This indecent assault seemed not quite right

To the bear who trotted off in a huff
On reaching the ground. **The smeared swain,** *tough*
Though he was, thanked his God with sobbing breath
For wondrous escape from untimely death.

25 Butler, Ch10, para 13.

26 Pliny, Book XI Ch 14.

27 My poem expands on Butler's account in Ch 6, 51; words not in italics are his.

Russia's Bashkiria region has a long tradition of managing both forest and hive bees and claims to produce a very superior honey.

BASHKIRIAN HONEY – ВАШКИРСКИИ МЁД

This small lathe-turned pot, chestnut brown, incised
As though made with staves of oak, inscribed
In Cyrillic, holds exotic smells, another world.

Stout women in floral headscarves crouched
Round a spring with plastic flagons – greeting
Friends and this stranger as sweetly
As the waters singing up from the ground.
We gathered fungi and berries, crossed brooks
On rotting mossy trunks quietly alive
With woodlice, beetles and damp toadstool smells.

Tatiana gave me this honey, from their prized
Burzyan bees, foraging on fireweed
And grasses on mountain steppe,
Humming in wild hives harvested with knives
Among ancient forest limes and oaks
Or from apiaries bright as paint boxes
Beside the Belaya river, where wives, brown
As old combs, sit on split-log steps to fill
Wooden pots with molten sun under the eyes
Of imperial eagles and rare black storks.

Back home near a bus stop and a shop,
The pot fits my hand. I savour its complex
Aromatics, sweet and herbal, primeval.

Fig. 12-2 : Honey in variety at the National Honey Show, 1985

Courtesy of Paul Embden

Perhaps every country claims its ambrosia supreme. After tasting Tenerife's white broom honey, reputedly better even than fabled Hymettus, Brown[28] hymns his Australian bloodwood product, gathered in perfumed mountain air. New Zealand beekeepers boast of manuka's excellent medicinal qualities.

Modern Britain offers many choices of garden, thyme and heather honeys but our bees love the sugary nectar flowing from ubiquitous fields of gaudy yellow rape. Beekeepers are less impressed, as the crop distracts nurse bees from more urgent duties and the honey sets so quickly that combs have to be spun out unseasonably early.

Emily Dickinson's laconic poem offers a refreshing perspective:

> The pedigree of honey
> Does not concern the bee;
> A clover, any time, to him
> Is aristocracy. [29]

Broaching the Treasures of the Honey-House

Clean handling is crucial and some writers make further stipulations. Virgil commands a well-rinsed mouth and Pliny declares bees abhor thieving or menstruating honey gatherers.[30]

Atticans lifted their bounty as wild figs ripened but the heavens guided the Romans. For his two harvests, Virgil watched for the Pleiads 'springing upward, spurn[ing] the briny seas', and then for their plunging avoidance of the Watery Scorpion's stormy train.[31] Pliny's timetable is exacting: thirty days after a swarm, at full moon, in fine weather after the summer solstice when Sirius shines brightest.[32] But whether to harvest the spring abundance or leave it to build up a strong swarm is a delicate judgement. The bees will pine away or flee if too much is taken; too little and they become lazy. His last harvest is around the ides of November (13th) but then all their bee bread and most of the honey must be left to see them through winter.

Skeps were hefted to judge their weight. Warder harvests the lightest and heaviest, the first being doomed anyway and the other perhaps so full of brood they would eat his profits.[33] Occupants were generally killed over burning sulphur before taking their combs but Butler merely dopes them with smoke and wipes them off with a feather;[34] a practice that still serves. Hive improvements allowed separation of the bees without harm; Thorley introduced a tin plate between them and the full combs, so 'you have an hive or a box of honey, and all your bees saved…'[35] Modern clearer boards and bee escapes further refine his method.

Virgin honeycomb must be sifted; a hairy bag serves Butler, ridding it of skadons (grubs) and skandarak (pollen). Levett wrings the combs through a thin cloth into an earthen pan before hanging them to drain in ways familiar to any bramble-jelly maker. Modern spinners are quicker but the honey still needs straining, although some producers claim that the bits of wax, propolis and bees' legs add to its virtues.

Finally, the honey must be closed in dry pots.[36] Such traditional care may justify Harrison's ancient boast. Saleable honey must be scrupulously clean and opportunist bee robbers must not reach spillages and risk mayhem. Afterwards, and at a safe distance, the spun frames and tools can be left for them to clean up.

28 Herbert Brown (1923) *A Bee Melody*, p201.

29 *Collected Poems*, Book III. XII.

30 Book XI and Book XXI.46.

31 From *Georgics IV*, Dryden and Greenough translations.

32 *The Natural History* Book XI Ch 14.

33 Joseph Warder (1722) *The True Amazons*, p75.

34 Butler, Ch 10, Pt 2 para 8.

35 Thorley, p195.

36 Levett, p50.

(Clockwise from top)

Fig. 12-3 : Smoking bees and collecting honey. Copperplate by Christoph Weigel (1654-1725) in W.H von Hohberg's *Georgica Curiosa*. Staatlische Museen ©*bpk, Berlin.*

Fig. 12-4 : Uncapping a comb in a modern frame

Fig. 12-5 : Spinning out the honey with a modern extractor

Fig. 12-6 : Sieving the spun honey to remove wax particles and the occasional bee's leg

Courtesy of Paul Embden

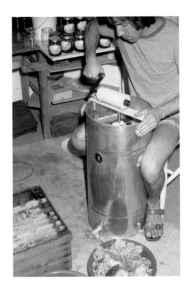

'AND IS THERE HONEY STILL FOR TEA?'[37]

'Bees make honey and men eat it', says an ancient Chinese proverb. Said to eat nothing else, the shadowy Homer composed his tales of the gods' duplicity and men's flawed virtues. Long lives were often attributed to the same narrow diet; when not writing theorems to torment our schooldays, Pythagoras spent his days contemplating music and souls, bee-like in transmigration. So important is honey in the diet of Congo pygmies, that they will abandon an animal hunt on discovering a bee nest.[38]

Honey cakes recur frequently in classical literature, and it remained the primary sweetener for centuries even after the cane sugar trade emerged around 700 AD. The Domesday Book reports honeyed porridge as an English staple. It was a valuable by-product of monasteries' candle-making activities until the Tudor Dissolution and Reformation drastically reduced both the ritual need and the number of apiaries. Lacking honey's flavour and abounding medicinal virtues, Butler deems sugar a poor substitute in all sorts of cooking.[39] Nearly two centuries later, a traveller in Greece records his simple delight in a saintly old priest's hospitality, better than any European princely banquet:

> *… delicate white honey combs, with bread and olives, and very good wine; to which he sat us down in his hut, and made us a dinner…*[40]

Honey now features little in standard English cookery; in the thousand-odd pages of *Mrs Beeton's Book of Household Management* it is mentioned just once and I rarely found it among modern recipes, and then only to supplement sugar. But beekeepers' associations and websites now promote its use. A basic recipe for honey cake requires:

> *3 oz butter*
> *3 oz sugar*
> *2 beaten eggs*
> *4 oz honey*
> *8 oz self-raising flour*

Cream the butter and sugar together and then gradually mix in the eggs. Beat well. Add the honey and mix thoroughly. Fold in the sieved flour. Put the mixture in a greased and lined seven-inch tin and bake for approximately one hour at 180°C/Fan 160°C/Gas 4.

The use of honey seems more prominent in central European cuisine – for example in gingerbreads and spice cakes such as *Lebküchen*. The now familiar *baklava* probably originated in Turkic Central Asia and versions of it spread throughout the Ottoman empire. Basically, filo pastry is layered with chopped nuts, spice and butter, then baked before being dowsed in a syrup of honey, sugar, water and vanilla. The Sultan was said to give these sweetmeats, made in Istanbul's Topkapi Palace kitchens, to his Janissaries on the fifteenth day of Ramadan.

37 Rupert Brooke *The Old Vicarage, Grantchester.*

38 Eva Crane (1983) *The Archaeology of Beekeeping*, p447.

39 Butler, Ch 10 Pt 3 paras 1-5, 16.

40 Mr Wheeler, quoted by Thomas Wildman (1768) *Treatise on the Management of Bees*, p97.

MARKETING

Late eighteenth century London saw a cut-throat trade in bee products. Thorley's small ads in the London papers boast of his:

> *… finest Minorca and Narbone [sic] Honey, in jars from two to 12 lb each, with all sorts of English and Foreign Honey; also the finest Honey in the Comb [and] Fine Cypress Honey… an excellent preserver of the lungs, in pots of all sizes.*

Among many other delights, including 'fine wild Thyme Honey, collected from all the herbmatic flowers', a suspicious rival announces:

> *MINORCA HONEY JUST ARRIVED, AT HOY'S Honey-warehouse, opposite Burlington-house, Piccadilly… N.B. To prevent mistakes, the Bee-Hive is over the door, and his pots marked Hoy, Piccadilly.*

Meanwhile, Scotland's prodigious harvests obviated the need for allegedly inferior imports. Next, Nutt asserts that one of his hives per square mile would, at one shilling per pound, profit 'an industrious, prosperous and virtuous peasantry',[41] except that imports undercut local trade.

Today's supermarkets and delicatessens offer a bewildering variety of sources and styles, each asserting unique qualities. Connoisseurs may choose acacia's pale, fluid gold and delicate flavour or manuka's strong, dark and creamier texture. Commercial blends demand consistency regardless of source or season, a mix determined daily by tasters and chemical tests. Yemen's admirable product in sticky plastic bottles[42] offers a striking contrast to Britain's rigorous hygiene standards.

THE HONEY PACKING PLANT

I certify I'm clear of infections,
Don steel-toed, non-slip shoes, an overall
And hair net, abandon my watch, my pen
And beads, sterilise my hands
Under a knee-activated tap. Then
And only then may my guided tour start.

***

Bees would circumnavigate
The Earth three times, it's said, to fill
One jar of honey. This yard is full
Of steel drums from all round the globe,
Each holding six thousand six hundred pounds.

41 Thomas Nutt (1832)
Humanity to Honeybees, p282.

42 Personal communication from travel writer Christine Osborne.

But we've hardly begun. Fifty thousand
Drums a year and polythene butts as big
As builders' cubic sacks disembark
On our shores, all freight
For one small town industrial estate.

I recollect familiar images —
Slaves to sunshine, bees cruising
From egg-yolk glow of crocuses
To ivy's autumn reticence. Endless
Murmuring and memories

Of lazy teas, dust motes in the air,
Sun loungers on the lawn.
All beekeepers know this timeless hum —
But this huge shed clangs and clatters
As jars are fed through a machine

To blow away any speck of dust
While fork lift trucks ghost past
To dump drums in hot cabinets which drain
Into troughs feeding shining risers,
First stages of a massive chemistry set.

The odd bee's leg or remnant of wax
Is filtered next, then unmarred gold
Pours into vats precisely warmed to fix
The crystalline consistency, varied daily
According to forage source and market choice.

The manager dips a thoughtful finger, tastes
And says there's much still to learn of honey.
We walk steel alleyways between the vats
And pipes, descend steel stairs to carousels
Where every jar receives its measured dose

Then marches to a capping machine,
Spins neatly past a labelling spool,
Arrives in human hands at last.
People unload the line, place the jars
On six-pack trays. Shrink-wrapped

And palleted, picked up by forklifts,
The finished products meet the air again.
Vans large and small wait open-mouthed,
Heading for massive distribution hubs
Or premium retail shops.

Sweet Salve, Milk of Paradise

Between mythic Olympus and the bottling plant is a transitional space where mankind receives and relishes the bees' ambrosial gift. Coleridge's ecstatic poem *Kubla Khan*, though more probably inspired by laudanum, ends with a warning.

> *He on honeydew hath fed*
> *And drunk the milk of Paradise.*

In *The Iliad* honey is often an actual or metaphorical pacifier among quarrelling generals. Wise old Nestor tries to reconcile furious Achilles and Agamemnon with 'words sweeter than honey'; honeyed bread and wine from his magnificent gold cup refresh exhausted fighters. Mourning beside Patroclus's byre, Achilles wishes an end to honey-sweet anger and battle while Athena cheers him with nectar and ambrosia. In *The Odyssey*, it comforts or tempts the hero at various points in his ten-year wanderings back to Ithaca. Aristophanes' anti-hero, Dikaiopolis in *The Acharnians*, simply wants peace, honeyed tripe and cuttlefish. But the Greeks had no monopoly over honey dreams.

The Bible promised ancient Israelites a land flowing with milk and honey (probably a fruit syrup); whilst wild bee honey, as eaten by Samson and Jonathan, was a rare delicacy. Later the Talmud agonised about the real stuff; insects are generally unclean but honey is acceptable because bees' bodies are seen as mere carriers. Its presence on the Seder plate celebrates the exodus from Egypt.

A Few Stories

Asenath, an Egyptian woman briefly mentioned in the Book of Genesis, Chapter 41, and some apocryphal texts, was fleshed out in a sixth century legend. A profoundly anachronistic mediaeval version then weaves a rich fabric of virginity and eroticism,[43] now versified by me.

> HONEY FOR ASENATH
>
> *She was a kohl-eyed giddy girl*
> *With shining hair, watching the world*
> *From her tower window, spurning*
> *Every suitor Potiphar proposed,*
>
> *Even Joseph, a linchpin now*
> *Of the Pharaoh's court. A shepherd*
> *From Israel is no match for me,*
> *She said — then saw him at the gate,*
>
> *Commanding in his chariot*
> *Of gold drawn by four white horses,*
> *Himself clad in purple and gold*
> *And white, wearing a jewelled wreath,*

43 E.W. Brooks (1918) *Joseph and Asenath*.

Bearing a jewelled royal mace.
Besotted — she comes down the stair
Adorned in her finest array,
Rubies and gold embroidered robes.

Joseph, glutted by women's swoons,
Is unmoved until her father speaks,
No other man has touched her.
Thus aroused, Joseph says, My law

Forbids me to kiss a heathen,
So I'll pray for your conversion
To the Christian faith. *She returns*
To her tower, casts off her glories.

Tossing her robes from the window
For the poor and her sumptuous food
For the dogs, she weeps seven days
In sackcloth and ashes. Angry

Parents reject her but she prays
That Joseph's great merciful God
Perhaps forgives her ignorance
And evil ways. Archangel Michael

Appearing with the morning star
Looking like Joseph transfigured
By burning light, commands, Asenath,
Stand up; girdle your loins and breasts

In white. God loves a penitent,
Your name is in the book of life.
Put on your choicest trousseau robe,
Make ready for your marriage day.

Come, angel, *she replies,* sit on my bed
Where no man ever sat before.
Drink and eat of my wine and bread.
Michael asks for honey. Dismayed,

Asenath calls a servant boy
To run and fetch some from the farm
But You'll find some in your larder,
Says the angel. And so it was!

A comb white like snow and honey
Heaven's dew from your mouth, *she says;*
A cleansing, healing ointment
After her week of penitence.

His hands spark like smelted iron,
Giving her tongue the saving food
And promises — Flowers of life from
God's fountain, a blessed future

Fine as cedars in Paradise;
And you will be holy mother
Saving all who kneel in homage
Before God's everlasting throne.

A cloud of stingless snow-white bees,[44]
Purple, crimson and scarlet winged,
Now swarm around to build their comb
On her lips until he says, Begone!

And himself in a burst of fire
And lightning leaves for holy heaven
In his four-horse golden chariot.
While she descends the stair to marry.

Ireland had its own legends and I have freely adapted one from the ninth century.

MIDHIR'S INVITATION TO THE EARTHLY PARADISE

Fair woman, come with me to fairy land
And rejoice your soul in sweet melodies
From sunrise to moonset, where blackbirds' eggs
Delight your eye with fragile blue, where moors
Are purple-draped and soft waves lap the strand.

My people, sinless as snow, with hair fine
As primrose down, their eyebrows ebony
And cheeks foxglove pink, never see
Their children die. Our meadows are watered
By crystal streams. There is no mine nor thine,

A land beyond purchase with any money.
Fair woman, come now! Your crown shall be gold,
Your gown of spider silk woven with flowers;
Your thirst shall be slaked with the finest mead
And your dish be filled with limpid honey.

44 Apparently white bees do exist in the Black Sea area – but surely not with such gorgeous wings!

Honey even features obliquely in that supreme cosmic drama, when God threw the rebellious angels from Heaven and expelled mankind from Eden. The angels were unconditionally blamed, Milton writes, but merciful God sought a volunteer to pay mankind's penalty; when his Son stepped forward:

> *... ambrosial fragrance fill'd*
> *All Heaven, and in the blessed Spirits elect*
> *Sense of new joy ineffable diffus'd:*
> *Beyond compare the Son of God was seen*
> *Most glorious...*[45]

The Finnish *Kalevala*[46] is a medley of traditional Creation myths, love stories, destructive monsters and terrifying nature. One sequence tells of Ahti (Lemminkäinen or Lover-Man); like a fairy story it encompasses universal themes — of three heroic trials, death, resurrection, the seasons, an indomitable mother and the agency of bees. The comely but arrogant young man ventures out to win the Maid of the North. Overcoming fearsome challenges, he finally meets a hideous end in Lord Tuoni's dark river of Death. Sensing her son is in trouble, his doting mother obtains an enormous rake from the smith Ilmarinen and gathers up all his shattered parts. Even when sewn together by many good spirits rowing boats up and down his veins, she still cannot revive him. It is only after a bee's three quests for honey that he is finally restored. The first cargo, gleaned in Tapiola, was not enough; neither were thirteen thumb-sized potfuls from beyond nine seas. On the third occasion, the mother begged her:

> *O bee, bird of the air*
> *fly there a third time*
> *high up into heaven*
> *above nine heavens....*[47]

So the bee soared beyond the Great Bear and returned with a hundred hornfuls of charmed honey from the great God Ukko and a thousand other ointments. Rubbing the salve into his patched-up body, his mother calls Lemminkäinen back from death and, scolding, takes him home to his wife.

HONEYED WORDS

Nurturing the Greek gods' abandoned bastard offspring, bee nymphs were the original Muses and bees were their Birds, thus involved with notions of wisdom and prophecy. When landing on human lips they bestowed gifts of oratory.

Pindar lived during Classical Athens' glory period. A bee settles on his sleeping childish form, Pausanias says, to gather honey from his lips. Other stories tell of him dreaming his mouth fills with honey and of Pan singing his poems on the mountains. Among many political and public commissions, only his odes to victors at the Games have survived intact.[48] They all start with praise for the winner, then hymn mythic heroes and the presiding gods and oracle of the particular Games.

45 John Milton (1667)
Paradise Lost, Book 3
lines 135-9.

46 Lönnrot's nineteenth
century epic poem.

47 Excerpt from Canto 15,
transl. Keith Bosley (1989).

48 Pindar Odes, *Pythian* 10.54;
Nemean 7.56; *Olympian* 6.

So for instance, at the Pythian Games, charioteer Arkesilas hears how the Delphic Bee gave glory to Jason's descendants and in particular to him. Proud parents are told to feast on brimming honeycomb while holiday songs dart bee-like from one thought to another! The boy Aigina, a multiple winner, is warned that excess, even of honey, cloys. At Olympia, part of Ode Six tells of Iamos, a love-child of the gods, abandoned by his shame-filled mother in an iris bed. There he is raised on honey by two bright-eyed serpents before visiting Mount Olympus to retrieve his birthright and immortal name Iris.

A swarm having settled in his cradle, Plato becomes known as 'the Athenian bee', blessed with wisdom. Though not bee-blessed in infancy, others are given the tag, like Sophocles 'the Attic bee' whose tragedies spoke with honeyed eloquence. Aristophanes creates a fantasy of peace where birds cull ambrosial music.[49]

Jewish traditions also associate bees with wisdom and oratorical blessings. Two revered Biblical women[50] bear the name Deborah (bee in Hebrew). One nursed Rebekah, the patriarch Isaac's eventual wife; the second was a wise prophetess. Driving a honey-laden chariot and bidden to eat a heavenly scroll sweeter than best honey, the visionary Ezekiel is a rushing wind of divine eloquence. Folk customs (see Chapter 15) continue to link honey blessings with a child's future studiousness and sweetness.

Norse poetry, particularly rich in vigorous anecdote and metaphor or *kennings*, links honey, mead and wisdom in complex myths. This magic mead[51] fell as a gift into the human realm. One such beneficiary would be Einarr Skúlason, a thirteenth century poet who contributed to the Old Norse *Morkinskinna*. Prosaically the title means mouldy parchment but it chronicles the glorious and gory history of mediaeval kings, participants in many years of Norwegian civil wars.

Christians also have their share of bee-blessed scholars and orators foraging for nectar, including Saint Ambrose, who wrote hymns for the laity to sing. Fra Filippo

49 *The Birds.*

50 *Genesis* 35.8; *Judges* 4.

51 See Chapter 3.

Fig. 12-7 : The Miracle of Saint Ambrose, Staatlische Museen
©bpk, Berlin

Lippi depicted a wondering audience round the saint's cradle as bees emerged from his mouth but ironically an over-enthusiastic restorer erased them, thinking they were specks of dirt![52]

Sharing in divine wisdom, bees underpin music and bestow human eloquence, according to Butler.[53] It seems triply fitting that Ludovicus Vives, in 1520 appointed Professor of Oratory at Corpus Christi, the College of Bees, was welcomed by a swarm settling in his study and himself became known as the Mellifluous Doctor.

Compliments Returned

So much for bees' gifts to orators and scholars. In return, Marcus Valerius Martial declares 'Every epigram should be like a bee', having sting, honey and brevity. A simple saying, probably from the West Country, is charmingly free of sting – 'I'm as 'appy as a bee in a snumper!' (foxglove). Nineteenth century Rilke's poetic goal is tragically exalted; an Angel will not hear him foraging for the earth's fragile essence.

> *We are the bees of the invisible. We desperately plunder the honey of the visible to gather it into the great golden hive of the Invisible.* [54]

Let us leave the matter on a less heartbreakingly ambitious note and think of the frequency and ambivalence of bees and honey in love poetry, for instance John Cleveland's 'Fuscara, Or The Bee Errant', with its lady sweet enough to delude a bee. Yeats creates a dream world of Innisfree with its 'bee-loud glade' and Tennyson[55] invents an idyllic childhood:

> *The yellow–banded bees,*
> *Thro' half-open lattices*
> *Coming in the scented breeze,*
> *Fed thee, a child, lying alone,*
> *With whitest honey in fairy gardens cull'd —*
> *A glorious child, dreaming alone,*
> *In silk–soft folds, upon yielding down,*
> *With the hum of swarming bees*
> *Into dreaming slumber lulled.*

52 Personal communication from Dr Catherine Whistler, Ashmolean Museum, Oxford.

53 Butler, Ch 1 para 63.

54 Quoted from Rilke's letters by T. Curtis Hayward, (n.d.) *Bees of the Invisible: Creative Play and Divine Possession* (London: The Guild of Pastoral Psychology).

55 *Eleanore* ii.

13 : MEDICINAL AND OTHER 'SUBTIL GIFTS'

PHARMACOLOGY

Food of gods and man since time began,
The nymph Melissa showed it good to eat,
Wise folk prescribe its use as medicine.

Men raided caves and trees; what risks they ran
To carry home the luscious priceless sweet —
Food of gods and man since time began.

Clay tablets in ancient Sumerian
Record the oldest clinical receipts —
Magi prescribed its use as medicine.

Mead antedates the oldest wine
For sacrifice, feast and grief — bittersweet
Food of gods and man since time began.

Aristotle declared a doctor's skill lay less
In what than when to use herbs or heat —
He too prescribed its use as medicine.

Nectar and bee enzymes: and so much more —
Thyme and beans differ from meadowsweet —
Food of gods and man since time began.

Modern doctors, learned monks, the Qu'ran
Advise — for gripes, perhaps MRSA, frostbitten feet —
Wise folk prescribe its use as medicine.

Lift combs, slice off the caps, let it drain —
Pot it to eat, mix in ointment, meet
Food of gods and man since time began,
Wise folk prescribe its use as medicine.

THE OLDEST MEDICAL PRESCRIPTIONS

1 S.N. Kramer (1956) *From the Tablets of Sumer*, pp57-58, interpreted with the help of a historian of science. Also Daniel T. Potts (1997) *Mesopotamian Civilization: the material foundations*, p150.

A long and honourable history was followed by centuries of neglect but honey is now reclaiming a place in scientific enquiry and clinical practice. Among the most ancient writings, Sumerian clay tablets over 4000 years old reveal, among surprisingly complex medical procedures, honey prescriptions.[1] Over half the medical papyri from Pharaonic Egypt feature its antiseptic and obstetric use. In many later and diverse writings it appears as a universal panacea, though Greek Hippocrates,[2] father of medicine and famous for his medical oath,[3] warned it was no cure-all; that could earn it an unfortunate reputation for hastening death. Aristotle also reflects:

> *... it is easy to know what honey, wine and hellebore, cautery and surgery are, [but]*
> *to know how and to whom and when to apply them so as to effect a cure is no less*
> *an undertaking than to be a physician.* [4]

Hippocrates dilutes it as an expectorant, also to cause heat and cleanse various sores and carbuncles. Honey from nests in his sepulchre was particularly favoured to cure children.[5]

Pliny's antidote to the evil *Azalea pontica* (or 'goat's death') includes finest honey, wine, rue and an emetic of salt meat. Oleander honey from the same region, being irregularly poisonous, he considers one of Nature's dirty tricks.[6] Apparently understanding that flies pollute, he recommends honey medicines from fly-less Mount Carina in Cyprus.

The Talmud and Qu'ranic commentaries both make much of its medicinal properties. Having prescribed it for a man with violent belly pains whose impatient brother complained that the cure was not instant, Mohammed counselled patience – 'God speaks truth and his belly lied'.

EARLY MODERNS AND 'BAREFOOT DOCTORS'

Galen brought Hippocrates' ideas to Imperial Rome, infusing Dioscorides' *De Materia Medica*, an authoritative text until the seventeenth century. Thomas Mouffet retailed his warning against poisonous Pontic honey and dirty imports but the ancient recommendation of honey to promote good infant teeth would make a modern dentist wince.[7] With many Biblical and classical citations, Butler's catalogue of its medicinal effects offers a 'found poem'. Honey:

> *easily passes into the body,*
> *cleanses, opens obstructions,*
> *clears breath and lights of bad humours,*
> *looses the belly and purges foulness,*
> *provokes urine, looses phlegm,*
> *clears eye troubles, nourishes much,*
> *breeds good blood, warms, prolongs old age,*
> *disinfects and preserves; embalms,*
> *is a mouthwash and gargariser,*
> *helps griefs of the jaws;*
> *is good against the bites of serpents*
> *and mad dogs, and poisonous mushrooms;*
> *is good for falling sickness,*
> *and a remedy against a surfeit.* [8]

2 Hippocrates *Medical Works.* Treatises ascribed to him probably have composite authors.

3 The modern BMA version requires doctors 'to help the sick to the best of my ability and judgement… abstain from harming or wronging any man…' An unfortunately sexist wording for modern times!

4 Aristotle *Nichomachean Ethics*, para 1137a.

5 Johannes Jonstonus (1657) *A History of Wonderful Things* cites this from Augustinus Gallus.

6 Pliny *Natural History of Animals*, Book XXI Chs 44-5.

7 Wm Harrison (1587) *Elizabethan England* (1876 ed.) pp175-6.

8 Charles Butler (1609) *The Feminine Monarchie*, Ch 10, Pt 3 para 1.

Among yet more virtues, he quotes the prescription of an 'excellent Kymist' – gold dissolved in a quintessence of honey to revive the dying. Fifty years on, Purchas's roll-call is similar[9] but he mixes honey with cheese to cure looseness and colic or with vinegar to ease constipation. Moreover, it soothes earache, reduces baldness and kills lice. It preserves old folks' senses and cured a stone-blind horse. 'Because it is hot in the second degree, [it cures] cold agues and moyst stomacks' but is not for fevers, jaundice or young men! John Evelyn blends it with oak coals to cure carbuncles and with distilled walnut leaves and urine to make hair spring on bald heads. A spoonful night and morning is a panacea for asthmas, coughs and thick phlegm.

Bees themselves, crushed or burnt, were used in teas, ointments and honey pastes for equally diverse conditions; with diuretic wine, Topsell also adds they are mighty cures for 'the dropsie… and all spots and flecks in the face'.[10]

As sugar encroached as a sweetener, Nutt took the opportunity to bolster the medicinal market, boasting that his cooler hive design and more hygienic harvesting system made his product particularly suitable.[11] One reads with relief that pressure should prevent sting toxin circulating, rather than one J.H. Payne's five minutes from a live coal!

Modern day Yemenis and Tanzanian tribesmen use honey as diversely as our forefathers did and westerners never abandoned it entirely. As every grandmother knows, it is a routine ingredient in cough mixtures. It is also a pill sweetener. Vets and falconers use it as a pick-me-up for their weary charges. A century ago 'poor struggling mites' in the New Jersey Baby Hospital[12] thrived on a diet of honey and skimmed milk. In the Manitoba wilderness, beyond any doctor's reach, a nun was strikingly successful with a wanderer's severely frostbitten limb, already blackening and potentially needing amputation. Having treated her own mild frostbite with honey, she also saved his foot.[13]

Rudolf Steiner, extrapolating from similarities between the inner structure of bones and the honeycomb, once again pursues an eccentric argument. Our bodies being made of a kind of wax, nursing mothers should take honey to build up their children but elderly bones become brittle with excess of form if they eat too much.[14]

Investigating shamans and healers, American psychiatrist Doug Boyd[15] tells of Shoshone Rolling Thunder who saw bees as willing colleagues. He instructed an urban chemist to ask the bees to share the plants and promise them no harm. On the third plant:

> *… they all made the strangest buzzing sound… as though they were… telling me to stop, and I was understanding… Rolling Thunder [then] said, 'There now, you see? You and the bees have agreed to share and now you're cutting back too far. They'll expect you now to do as you said.' So I cut only the front half very carefully… And he said that this was a gift of the Great Spirit!*

NEW SCIENTIFIC AND CLINICAL INTEREST IN BEE PRODUCTS

While some old nostrums are implausible, experience has endorsed others and today's scientists are taking a fresh interest in honey's complex chemical properties

9 Samuel Purchas (1657) *A Theatre of Politicall Flying Insects*, Ch 6.

10 Edward Topsell (1658) *The History of Serpents*, p645.

11 Thomas Nutt (1832) *Humanity to Honeybees*, Ch 2.

12 A.I. Root (1908) *The ABC and XYZ of Bee Culture*.

13 Diana Cooper (1985) *Bee Tidbits*. www.chebucto.ns.ca/~ag151

14 Rudolf Steiner, (1998, transl. Thomas Braatz) *Bees*, p91.

15 Doug Boyd (1974) *Rolling Thunder*, Ch 14.

and the therapeutic potential of other bee products. I can only offer here a taste of the wealth of research information appearing on the internet.

Honey

Antibacterial and non-stick properties for dressings are well confirmed, drawing fluid to bathe and debride wounds. It also kills harmful gut bacteria and ameliorates irritable bowel syndrome. New Zealand's manuka honey makes particular claims to medicinal qualities, its calcium alginate being good for leg ulcers and various skin complaints.[16] *Beecraft* magazine recently reported encouraging work in Belfast on honey's versatile use against MRSA and the BBKA is canvassing research into the differential properties of our native honeys. Local honeycomb, complete with pollen and cappings, may immunise against hay fever and one company promotes the qualities of honey gathered from flower-rich Derbyshire countryside.[17]

Several German airports now recruit bees as colleagues because honey has proved a useful telltale of pollution. That 'Düsseldorf Natural Honey',[18] from airport hives, has consistently proved as pure as rural honey is a good mark for European air quality regulations.

Mead

Butler claimed versatile therapeutic virtues for this most agreeable wine; it 'restores appetite, opens the passages for the Spirit and breath, and softens the bellies',[19] also relieves coughs, ague and the falling evil [epilepsy]. Since it prevents general debility and promotes longevity, he attributes the ancient Britons' fair complexions and sound bodies to their copious consumption of it.

Bee venom

Used to treat rheumatism for thousands of years, it was scientifically investigated in late nineteenth century Austria. Longgood[20] reports a strange testimonial; bees attracted to a bullied chicken stung it over 200 times before it was rescued, whereupon it recovered overnight and thereafter unprecedentedly laid an egg every day! Sporadic reports of medical successes appeared until in 2004 Korean researchers located the effective ingredient, mellitin.[21] The venom is now officially licensed in Beijing and to British beekeepers to immunise themselves against anaphylaxis. Preliminary Australian research[22] points to cancer treatments possibly less toxic than those currently available.

Wax

Purchas recommended it as an ointment for chapped lips and nipples, and to 'mollify sinews'. More questionably, he said, taken internally it prevents breast milk curdling and a poultice of it would cure worms.[23]

Propolis, the hive sealant

Aristotle knew of its powerful antiseptic properties and Florence Nightingale valued them. Currently a major airline is exploring a propolis mist to sterilise aircraft. Even Pliny's ancient claim that it cures tumours is receiving renewed attention, with Japanese scientists finding potent anti-oxidant and tumoricidal effects. An Italian team[24] is investigating other properties including liver protective functions.[25]

16 Advancis Medical Ltd. www.medicalhoney.com

17 Medibee Ltd. www.medibee.co.uk

18 Various websites including www.scholastic.com/browse/article.jsp?id=3754880 'Biodetective Bees', 13th September 2010.

19 Butler, Ch 10, 21.

20 William Longgood (1985) *The Queen Must Die, and Other Affairs of Bees and Men*, p180.

21 Several papers available on the internet, e.g. in *Arthritis and Rheumatism*, November 2004, 50.11.

22 CSIRO Australia (1999, March 4). Bees – Latest Weapon In Cancer Fight.

23 Purchas, Ch6.

24 Stefano Castaldo and Francesco Capasso (2002), 'Propolis, an old remedy used in modern medicine', in *Fitoterapia* November Suppl. 1:S1-6. Abstract on PubMed.gov.

25 For instance, Y.Nakajima et al (2009) 'Comparison of bee products based on assays of antioxidant capacities', in BMC Complementary and Alternative Medicine 9.4. Numerous other web references.

LOTIONS AND POTIONS

Ever since Cleopatra bathed in honey and ass's milk and Nero's second wife, Poppaea, perhaps more sparingly, used the same as face lotion, the beauty industry has eulogised bee products. Italian and French aristocrats used them in various ointments and they are common in modern moisturisers. Extravagant claims are made for beeswax in lipsticks and face creams. But this is Roman Aldrovandus's curious recipe for a man's toilette – 'If thou wouldst have thy beard grow quickly, anoint thy chin with the ashes of burnt Bees, and Mice dung.'[26]

WASSAIL![27]

A German peat bog revealed a two-thousand-year-old mead drinking horn, the oldest such artefact, but many myths vouch for the drink's timeless status: Indra's eagle; Zeus's eagle; Odin's shape-shifting quest. People would have enjoyed the happy accident of water falling into honey and causing rapid fermentation long before they learned to make wine.

A fairy horse takes the Irish lovers Ossian and Niamh to a heaven flowing with rivers of mead. A princess refreshes the hero, Finn MacCooll, with a silver cupful and divine maidens in Valhalla reward fallen warriors with it. Finns thought the heavens contained a honey store and the Chinese found it brought heavenly pleasure. A Polish legend[28] tells of poor wheelwright Piast, after his hospitality to persecuted strangers, being rewarded with mead-packed cellars and full granaries (plus the throne for his son!).

A vital lubricant in social gatherings, mead flowed freely among Britons and western Celts, according to Pliny and Strabo.[29] Tara (home of the High Kings of Ireland) is known as 'the House of the Mead Circle'. In a miracle echoing the marriage at Cana,[30] Saint Brigid saved a relation's blushes by filling his empty vessels with mead before a king's visit. The sixth century Welsh poet Taliesin sings of its blessed abundance and in the *Mabinogion*'s story 'Owain', for instance, King Arthur invites his guests to share a flagon and exchange tales of their adventures while he has a nap.

'*Wacht heil!*' '*Drinc heil!*' ('Your health!') Ancient Germanic[31] traditions of roistering companionship thrived on plentiful supplies. Profane notes are struck by Odin's goat whose udders eternally dispense the liquor and so, too, by Thor's feat of once drinking two tuns. Beowulf's mead-hall is the greatest ever known, where people laugh and:

> *… nor in all my life saw I*
> *Under heaven's vault, among sitters in hall,*
> *More joy in their mead.* [32]

The Ynglinga Saga tells earthy tales from shadowy times, of endless berserking bouts of war between men with ruffianly names. Descended from great King Odin, they harried far across land and sea to feed 'the falcon of the weapon wave'. But in peaceful interludes these massive men displayed massive appetites for mead[33] and friendship, which could lead to actual misfortune or perhaps to fictional anecdote as my poem elaborates below.[34]

26 Jonstonus, p245.

27 G.R. Gayre (1948) *Wassail! In Mazers of Mead.*

28 Derived by Gayre from Monica Garland (1942) *Poland.*

29 Strabo (d.23 AD) *Geographia*, Bk IV,v.5.

30 *Gospel of St John*, Ch. 2.

31 Interestingly, the modern German word '*bier*' derives from the same source as '*bienen*', bees.

32 C.K. Scott Moncrieff (1921) *Beowulf*, lines 1015-7.

33 Though by the time Snorri wrote, beer was the drink of choice.

34 From Snorri Sturlusson's *Heimskringla.*

BEWARE EXTRAVAGANT HOSPITALITY

King Fjölne sailed from Sweden to Selund
To visit Fróthi whose grand hall housed
A mead bowl of prodigious size and weight

Many ells in height and built of ship timbers,
Such that no man could lift it or reach in.
It must be filled and ladled from above

Through a hole cut in the upper floor. Churls
Ran up and down the stairs, filling tankards
Uncounted times while others loaded plates.

The bellies of kings and thanes filled with meat
And their heads with wild tales and laughter.
The evening offered hours of fellowship

And Fróthi's mead was famously strong.
They drank as keenly as they fought.
At last they were assisted to their beds.

In the darkest hour, Fjölne was overcome
By his bladder's necessities. He stumbled forth
To find the privy in this unfamiliar house,

Tripped on a roughly knotted board
And plunged headlong through the floor
To drown in the bowl of mead.

So sadly ended a visit of goodwill, leaving
New centuries open for feuding princes,
Warlords over petty provinces

And marauding Vikings on the seas,
With names like Ragnar Hairy-Breeks
Or Erik Bloodaxe to terrify the peasantry.

More solemnly, the Venerable Bede likens human life to a sparrow's brief flight through the mead hall, where you sit:

> *… feasting with your ealdormen and thegns… In that time in which it is indoors it is indeed not touched by the fury of the winter, but yet, this smallest space of calmness being passed almost in a flash, from winter going into winter again, it is lost to your eyes…*[35]

Mediaeval monks drank mead freely on feast days but were forbidden to mix honey with (base) beer, perhaps because of ancient sacral usage. Chaucer's[36] characters drink it: in the Miller's Tale the carpenter's lovely young wife has a mouth 'sweet as

35 *History of the English Church and People*, Book II Ch 13.

36 David Wright's translation of *Canterbury Tales*, pp83,1701-2.

mead'; elsewhere the comic spindly knight Sir Topaz, after narrowly escaping Sir Elephant, fortifies himself for further battle.

The drink remained important until farming displaced forests and their bees. Then the dissolution of the monasteries shrank English honey production further and mead gradually disappeared as wine and beer became relatively cheaper and kept better. However, the spiced variant metheglin lingered on. Shakespeare's characters drink it and in 1666 Pepys enjoyed a dinner of delicious royal left-overs, that included:

> *… a most brave drink cooled in ice… metheglin for the King's own drinking which did please me mightily.*

In due course, bee chandlers like Thorley boasted improved recipes, as good as the best of foreign wines, and Pasteur's discoveries made it easier to keep it palatable. Even so, wine won out.

Beekeeping, home brew, health food circles and country fairs have promoted some revival in more recent times. Paying rental to farmers for his hives with Buckfast Mead, Brother Adam led the way. Maturing his best in oak sherry casks for five years, he describes four types, from champagne-like to spiced metheglin, and his book includes scrupulous instructions for making them.[37]

This Anglo-Saxon riddle irresistibly suggests modern Saturday nights on the town:

> *I am cherished by men, found far and wide, brought from the groves and from the city-heights, from the dales and from the downs. By day wings bore me in the air, carried me with skill under the shelter of the roof. Afterwards men bathed me in a tub. Now I am a binder and a scourger; straightway I cast a young man to the earth, sometimes an old churl. Straightway he who grapples with me and struggles against my strength discovers that he must needs seek the earth with his back, if he forsakes not his folly ere that. Deprived of strength, doughty in speech, robbed of might, he has no rule over his mind, feet or hands. And what is my name, who thus on the earth in daylight bind youths, rash after blows.*[38]

37 Brother Adam (1987) *Beekeeping at Buckfast.*

38 From R.K. Gordon (1942) *Anglo-Saxon Poetry*, p330. © Everyman's Library.

39 Australian television documentary 'First Footprint' by Martin Butler and Bentley Dean, narrated by Ernie Dingo, October 2013.

40 J. Frith *et al* 'Sweetness and Light: Chemical Evidence of Beeswax and Tallow Candles at Fountains Abbey, North Yorkshire', in ads.ahds. ac.uk *Notes and News* vol48, 220-227.

41 John Levett (1634) *The Ordering of Bees*, p51.

WAX – BLESSED AND PROFITABLE LIGHT

Besides creating bees' marvellous architecture and reputedly curing various ailments, wax has significant economic and industrial value. Over 40,000 years ago, Australian Aboriginals used it with honey to glue their stone tools to wooden shafts.[39] Romans used framed tablets as erasable notepads and wax candles for his chamber featured in a mediaeval priest's retirement pension.[40]

Methods of reclamation hardly change; once all the honey and detritus have been filtered, any unwanted comb, heated and water-cooled, is moulded as you please, as Levett advises.[41] Modern practice is to keep the cleaned scraps in moth-proof tins until there is enough to make bricks proudly marked BEESWAX.

An English Act in 1580 decreed the purity of traded wax. Purchas is astonished that Russia could annually export a thousand tons and still use it prodigally to light great men's houses. China, too, has 'Much waxe, you may lade… fleets therewith'.[42] In England it becomes the lighting of choice 'in all polite assemblies; as well as in the Romish churches',[43] its fragrant and clean-burning providing a seemingly cut-throat commercial opportunity. Warning against inferior German products and hawkers selling fat merely clad in wax, Thorley advertises his prompt, twice daily deliveries to the West End, five shillings cheaper than shop price, of:

> … *real HAMBRO CANDLES… Virgin Wax Candles guaranteed against guttering [and preferred by] most of the Nobility and persons of fashion…*[44]

Nowadays, such candles are luxury items for boudoirs and bathrooms. Hans Christian Andersen's story, *The Candles*, contrasts a boastful beeswax item living in a chandelier whilst the tallow candle is content to light the kitchen.

But wax has other uses, noble and prosaic. It embalms the dead and sometimes serves the newborn; in Vietnam Crane saw baby slings of most intricate batik, striped in different colours, the resist very finely applied with crushed plant stems.[45]

It contributes to polishes, lubricants, threads and waterproofing. Among forty current industrial uses are in armaments, electrical transducers and fine moulds for metal castings. Accidental thumb-prints preserved on a Bronze Age tripod moved Edwardes[46] and provoked me to this poem.

THE TRIPOD ARTIST

The man will be for ever nameless
But on an ancient tripod we can trace
The faint imprint of his sweaty hand.

This Cypriot, skilled artist, depicted
A frantic chase. To bring their people meat,
Hounds lunge after wild goats in flight.

First shaped on malleable, transient wax,
His fluid lines record their common life —
The hunt in action, the ceremonial meal.

A cooking pot to make dinner for the rich
Has immortalised his mark —
Brother, husband, father, artisan.

42 Purchas, p140.

43 Thomas Wildman (1770 ed.) T*reatise on the Management of Bees*, Preface.

44 Thorley, John (1st ed. 1744; 4th 1774) *Meliselogia or the Female Monarchy*.

45 Eva Crane (2003) *Making a Beeline*, Ch 14, p196.

46 Edward Edwardes (1908) *The Lore of the Honeybee*, p.xii.

A Tragic Myth

Daedalus's fine modelling of a gold honeycomb was noted in Chapter Ten but perhaps Icarus is more widely memorialised. Wax failed its ultimate test when, defying his father, he flew too near the sun and his wings melted. Brueghel's painting[47] features just his legs disappearing under the waves while everybody about their daily business fails to notice the tragedy.

47 'The Fall of Icarus'. Apparently now considered to be a very good copy but not Brueghel's original.

Fig. 13-1 : Icarus

Courtesy of Andrew Gough's Arcadia. www. andrewgough.co.uk. Original source unknown

Fig. 13-2 : Pieter Brueghel the Elder – The Fall of Icarus (see footnote 47 opposite)

© Brussels Museum of Fine Arts

14 : COMPANIONS FROM CRADLE TO GRAVE

BEE BLESSINGS

Once again we find hazy boundaries between themes. On life's journey, folk customs and superstitions frequently contain marked religious features. Bees often feature in solemn rites of passage.[1]

BEE BLESSINGS

Folk pour blessings on every infant's lips —
Across all ages and all continents
Give spoonfuls of honey in gentle sips.

Young love forgets the rose has thorns; parents
With honey cakes hope the nectar of the bee
Will for ever flavour today's commitments

While each wedding guest toasts prosperity,
Clement weather for the honeymoon,
Good health, fine children and longevity.

When tolling bells spell out a solemn tune,
Embodying the souls of all the dead
Bees must be told their keeper's gone.

While still the mourners' tears are shed,
Honey bathed, the body waits to lie below
Until every final prayer is said.

Such sacred meanings do the bees bestow
On each and every delight and sorrow.

BIRTH AND CHILDHOOD

To the ancient Greeks, honey symbolised the pleasure that attracts souls to be born. In bee form and named *melissae*, virtuous souls sprang from the union of Artemis/Melissa and the horned bull/crescent moon.

Nectar-born and seated on lotus flowers, India's great gods are all Madhava.[2] Even before the umbilical cord is cut, Jains, Hindus and Muslims may all touch a baby's lips with honey and butter from a golden spoon, while the father whispers in the baby's ear for the gods' protection and a hundred autumns.[3]

1 Once again, I am indebted to Hilda M. Ransome (1937) *The Sacred Bee*, for several ideas and sources in this chapter.

2 Chapter 13 notes that the Sanskrit *madhu* means honey/sweetness; it forms the root of such titles as Madhava and gives deep significance to honey rituals.

3 N. Laroia and D. Sharma 'The religious and cultural bases for breastfeeding practices among the Hindus', in *Breastfeed Med.* 2006;1: 94-98.

4 Isaiah 7.14-15. Traditionally interpreted by Christians as the future Jesus Christ.

Given to a Jewish newborn, honey promises moral wisdom.[4] It enlightened and fortified Jonathan before battle (1 Samuel 14). A child would traditionally be given a slate of honey-covered letters to lick clean as each was learned and it is still customary for a child to lick a honeyed plate on starting school.

Such customs continued across a wide tract of Christendom.[5] In the name of God and the Virgin, a mother's nipples might be moistened with honey and wine or the baby's lips smeared even before its first suckling. This presumably encouraged the child to feed but also symbolised the hope for a long, happy life. In some countries it is an older child's job, wishing the newborn 'Be thou sweet as honey'. Galician Jews put comb or a honey cake in the cradle. A curious twist in Northern Europe, surely harking back to archaic associations between souls and bees, was that until it had tasted honey a baby was merely a 'soul', which could be killed.

SEALING THE SOUL

Maria, daughter, you've produced yet another brat,
Unwelcome proof of such fertility,
Opening our humble door to poverty.
I'm straight out with her to the washing vat

To drown as carelessly as a kitten litter
Before her soul acquires a title to this earth,
A foothold at our meagre hearth,
Or, warmed to her, your heart and mood turn bitter

Against me, your stooped, chap-handed mother.
I've raised seven of you and lost seven more.
When I was young I loved you each for sure
But now I'm done. Where is your no-good lover?

Good neighbour, why do you scurry
To the wash-house with that bloody sheet?
Maria's baby must have come. You should sit
With her, relaxed and slow. No hurry.

I've brought some honey to bless
The baby's lips and implant its flitful
Soul in its body-lodging. But that bundle
Holds, I see, the tiny, harmless

Scrap. Oh neighbour, you cannot mean
Her harm. Here, see, she grips my hand,
Sucks my honeyed finger; life claims this land
With all its pleasures, hardships, pain.

This sweet has sealed her place among us.
Take her to her mother's breast — you must.

5 A.E. Fife (1939) *The Concept of the sacredness of bees, honey and wax in Christian Popular Tradition.*

LOVE AND LUST

In the Linnaean classification bees are among the *hymenoptera*, literally veil-winged, a label serendipitously linking age-old bee cults and the broaching of a virgin's hymen. Reynolds painted the famous *Three Ladies Adorning a Term of Hymen*.

Veil-winged bees hide
The innermost shrine of the virgin goddess
From vulgar view; her priestess Hymen
Presides when young couples come
Joyously to wed, then dance and feast
Under the sun's bright blaze.

However, chastity and lust form a tangled skein. Dionysus discovered honey; Pan and Priapus are both protectors of hives. This modern account of a flower's response to the bee's touch is distinctly erotic:

Our humble country orchid will automatically incline and cause to adhere its pollinia; the wild sage will touch the visitor at a particular spot that the bee may touch the stigma of a neighbouring flower at a correspondingly vital spot... How profuse is nature in the gifts she rains on love![6]

'HONEY IN THEIR MOUTHS AND STINGS IN THEIR TAILS'

This French proverb expresses the fact that duality is everywhere in the customs, art and literature of love. From the sixth century BC, through chivalric romances and Eastern ghazals (a type of Persian poem) to the present day, poets pour forth streams of milk, maybe sour, and honey, often with a sting.

In Euripides' eponymous play, rejecting his step-mother Phaedra's advances, Hippolytus brings her a garland from Artemis's inviolate meadow where only bees may venture. Ravaged by guilt, she hangs herself and the chorus sings of love being like a destructive flitting bee. Who can forget, having once heard, Dido's lament in Berlioz's opera *The Trojans*? This is how I imagine the scene:

DIDO AND AENEAS

While builders' dust swirls around
The emerging golden grandeur
That will be Carthage, bees buzz
Unceasing, without demur.

The Trojan fleet waits at moorings
While Queen Dido and Aeneas
Pursue ecstatic love, their senses
Seduced by thyme and honey scents.

6 Herbert Brown (1923)
A Bee Melody, p110.

For a few brief, unmindful days
They, like bees, ignore winged time
And what the gods ordain. He will
Sail on to destiny and she

Will stand, abandoned on the quay,
Filling the air with tragic lament,
Then turn to mount a waiting pyre
Kindled by bee-nurtured shrubs.

Kama, the Indian god of love, carries a bow strung with a chain of bees. Eros/Cupid's propensity to sting dates back at least to Anacreon's ode written six centuries BC; stung while stealing honey he complains to his mother Aphrodite/Venus, earning the retort that he has caused many hearts greater pain. The theme so interested Lucas Cranach (or his customers) that he painted the image nine times.

Fig. 14-1 : Lucas Cranach (1525)
Cupid complaining to Venus

© *National Gallery Picture Library, London*

PANGS OF LOVE

Well-acquainted with the wiles of love
And the lee shores of the isles of desire,
Aphrodite would neither hug nor offer
Words of comfort to Eros running home.

For her sweet imp had thrust his hands and lips
Deep into a honeycomb. Busy, outraged
Virgins caught him, fingers dripping
With their sacred syrup, gold-handed,

And so they mobbed and stung until
He burned from head to toe and tears sprang
From cherubic eyes. His mother said, Now,
My pretty boy, you won't forget the arrow's pain.

An Irish proverb cautions against licking honey from a briar, and the Roman philosopher Boethius acknowledges the sting in the tail of pleasure thus:

> *Every delyt hath this, that it anguisheth hem with prikkes that usen it. It resembleth*
> *to thise flyinge flyes that we clepen [call] been, that, after that he hath shad hise*
> *agreeable honies, he fleeth awey, and stingeth the hertes, of him that ben y-smite,*
> *with bytinge overlonge holdinge.*[7]

In similar vein, the fourteenth century Sufi, Hafez of Isfahan, warns against the world's shallow rewards, for:

> *No-one ever gathered honey*
> *Without a sting or carried roses*
> *From your garden*
> *Without a scratch.*[8]

Bees fly through Elizabethan poetry, sometimes defending ladies' virtue with their sting, at others spreading lubricious honey. The Shakespeare canon covers all aspects of love but he is cynical about honey. Adolescent Hamlet is overwhelmed by his mother's lust for Claudius, accusing her of being 'stewed in corruption, honeying… / Over the nasty sty'[9]. Deflating Falstaff's delight in Mistress Quickly, 'a most sweet wench', Prince Harry ripostes, 'As the honey of Hybla, my old lad… and is not a buff jerkin a most sweet robe…?'[10] Elsewhere (in *The Tempest*) the asexual sprite Ariel sings 'Where the bee sucks, there suck I'.

Obliquely acknowledging Cupid, Thomas Lodge's 'Phillis', opens thus:

> *Love guards the roses of thy lips*
> *And flies about them like a bee;*
> *If I approach he forward skips,*
> *And if I kiss he stingeth me.*

7 Chaucer's *Boethius* Bk III, metre vii, from the sixth century *Consolations of Philosophy.*

8 *Mukathast,* my translation from a German source.

9 William Shakespeare, *Hamlet* 3.4.82.

10 William Shakespeare, *Henry IV Part 1,* 1.2.41.

Elsewhere, in a madrigal, the lady obviously enjoys succumbing:

> *Love in my bosom like a bee*
> *Doth suck his sweet;*
> *Now with his wings he plays with me,*
> *Now with his feet.*
> *Within mine eyes he makes his nest,*
> *His bed amidst my tender breast;*
> *My kisses are his daily feast,*
> *And yet he robs me of my rest.*
> *Ah wanton, will ye?*

The Elizabethan poet and friend of Shakespeare, Richard Barnfield, contrives an affectionate shepherd whose lady-love has a fruit-filled bower and an ambition to be 'thy hive, and thou my honey bee'. The Devon parson Robert Herrick plays the same tune. Besides reworking Anacreon's poem, another features two Cupids quarrelling about 'the sweet bag of a bee'. *The Captiv'd Bee; Or The Little Fiddler* describes a seduction scene with his sleeping Julia, enrolling honey very suggestively:

> *But when he felt he suck'd from thence*
> *Hony, and in the quintessence:*
> *He drank so much he scarce co'd stir…*

The poem concludes with her smiling and promising:

> *When next he came a pilfring so,*
> *He sho'd from her full lips derive,*
> *Hony enough to fill his hive.*

Variations carry on down the centuries. Burns declares that the bee's sips of nectar cannot compete with his 'delight… Upon the lips o' Philly'. Robert Browning, after inviting 'The moth's kiss first!' tentative and shy, boldly seeks in the second verse:

> *The bee's kiss now! Kiss me as if you entered gay*
> *My heart at some noonday.* [11]

Anxiety and guilt later replace such happy sensuality. In 'The Eve of St Agnes', only secretly in 'the honeyed middle of the night' does Keats's Madeline pine for love until her Porphyry penetrates the castle and carries her away. Tennyson's suitor of Maud likens love's cruel madness to 'the honey of poison-flowers'; his modest victory in meeting her in the garden (towards the end of a very long and rambling poem!) actually brings catastrophe. Another Tennysonian marathon tells a deeply ambivalent tale of Princess Ida,[12] a prototype liberated woman defeated by the tussle between love and equality; at last we find her in bed with her lover, reading him a pastoral romance about a shepherd piping his maid to a life of hearth and children. Her voice fills his ears, accompanied by sweet 'murmuring of innumerable bees'. In contrast, terse scepticism prickles through Emily Dickinson's 'Possession':

11 Robert Browning
'In a gondola'.

12 Alfred Lord Tennyson
'The Princess'.

Did the harebell loose her girdle
  To the lover bee,
Would the bee the harebell hallow
  Much as formerly?

Did the paradise, persuaded,
Yield her moat of pearl,
Would the Eden be an Eden,
Or the earl an earl?

Was it only men who sipped nectar? As the twentieth century dawned, Albert Fitz's popular song rejoices, 'You are my honey, honeysuckle, I am the bee'. Today's more permissive eroticism openly admits oral sex but are both parties equally gratified?

MARRIAGE

Among the earliest of all writings, this Sumerian love song shows the goddess Inanna delighting in her new husband, the king. Their union secures the world's fertility:

Bridegroom, dear to my heart,
Goodly is your beauty, honeysweet.
Lion, dear to my heart,
Goodly is your beauty, honeysweet...
My precious caress is more savory than honey...
Your place goodly as honey, pray lay (your) hand on it...[13]

Drowning in sensuous pleasure, King Solomon, too, sings:

Thy lips, O my bride, drop as the honeycomb;
Honey and milk are under thy tongue...
I have eaten my honeycomb with my honey.[14]

In Hinduism honey and mead are aphrodisiac and fertility boosters. Bridegrooms receive a cup of *madhuparka* (honey, curd and ghee) and pray for blessings, including honey-sweet nights. Honey spread over the bride wards off malign spirits and her husband anoints her with these words:

The speech of thy tongue is honey;
In my mouth lives honey,
And in my teeth lives peace.[15]

In contrast, Greek and African Thonga customs forbid women to eat honey for a year after marriage, lest they take off like foraging bees.[16] According to the Book of Proverbs (5.3), an adulteress's lips drip like honeycomb until her hell-bound end is bitter as wormwood.

Christian customs evidently incorporated older traditions; for instance, putting honey on a Polish bride's lips, while in Bulgaria a woman would rub honey on the

13 S.N. Kramer (1956)
From the Tablets of Sumer,
p250-1 (his own transcription and translation).

14 Song of Songs 4.11-12, 5.1.

15 Various Hindu websites.

16 Claire Preston (2006)
Bee, p126.

groom's face and command both to be 'fond of each other as the bees are fond of this honey'.[17] Many central European blessings are amalgamated in the following verse.

WEDDING PREPARATIONS

Mother bakes honey cakes
And father drapes the hives in red
For son or daughter about to wed.
Honey paints the groom's doorway,
A silver spoonful greets his bride.

Shower their bed with honey water,
The mop and broom and fire irons,
Their cutlery too and cooking pans,
To keep intact their love and home.

Wash her feet and sprinkle each guest,
Enjoy the wedding and the feast,
Quaff a cup of mead with the priest.
Then introduce the bees and seek
Their faithful help as children come.

LOSS AND SOLITUDE

In Pauline Stainer's poem 'The Honeycomb', a widower's memory of tremulous first loving blends with honey and bees humming in the bedroom wall. In contrast, Vita Sackville-West's widowed Lady Slane[18] relishes her freedom in old age:

For the first time in her life — no, for the first time since her marriage — she had nothing else to do. She could lie back against death and examine life. Meanwhile, the air was full of the sound of bees.

DEATH

Honey was a token of purity in the Sumerian ceremony of 'Opening or Washing the Mouth' – blessing the body during obsequies. Indians offer a rich honeyed oblation to Yama, the god of Death.[19] Mead features in Greek cults of the dead and in their mythic underworld. It continued in later libations, as when Medea tells Jason to pray before confronting flaming bulls and the field of dragons' teeth.[20] Exiled from Thebes after unwittingly killing his father, blind Oedipus[21] stumbles into the Furies' grove at Colonus and must placate them with three libations; the last bowl must be entirely drained of its honey. All such libations gratify the gods but also nourish the souls with their favourite foods and comfort the mourners.

Sometimes, like Keats, the living hanker after easeful death. After twenty weary years of war and arduous voyaging, Odysseus arrives home. Before confronting

17 Bee Wilson (2004)
The Hive.

18 Vita Sackville-West (1931)
All Passion Spent.

19 Ransome, p51.

20 Apollonius Rhodius,
Argonautica 3.

21 Sophocles *Oedipus*
at Colonus.

Penelope's tiresome suitors, he yearns for radiant Aphrodite's realm and a feast of sweet honey, cheese, and wine. Penelope, heartily fed up with the suitors and unaware he is home at last, simultaneously voices the same wish.[22] Since Aphrodite is the goddess of love, perhaps their yearnings were not so focused on an afterlife.

In a story with remarkable similarities to Snorri's,[23] Saxo Grammaticus tells of a Viking wake that went grotesquely amiss. A malicious plot to kill the Norwegian King Hadding was foiled but news of his death had leaked prematurely. His grieving Swedish brother-in-law, King Hunding, laid on a sumptuous feast with a liquor jar of extraordinary size and, as a token of respect, he himself acted as cupbearer. Everyone got riotously drunk until he stumbled and drowned in the mead. On hearing this, Hadding:

> … wished to pay like thanks to his worshipper, and, not enduring to survive his death, hanged himself…[24]

Purchas seems to have misread this knotty tale, declaring grief caused Hunding to drown himself deliberately '(but ridiculously and foolishly). Some applaud and prefer him therefore, before many heroical Greeks and Romans.'[25]

ANOINTING, EMBALMING AND COMFORTING THE DEAD

In *The Iliad*, after Patroclus dies in heroic battle with Hector, the goddess Thetis bathes him in ambrosia and nectar while promising her grief-stricken son, Achilles, 'that his body shall be as it was'.[26] Oil and honey jars surround his pyre – and Achilles' own in his turn – while Odysseus pours honey before Tiresias's soul guides him through death's perilous realms. After the goddess Artemis delivers Iphigenia from a sacrificial death, she appoints her to 'pour fragrant honey from tawny mountain bees' on the graves of men dying in the Crimea[27] and in Aeschylus's tragedy, *The Persians*, broken-hearted Queen Atossa brings it to soothe King Darius's ghost.[28]

Ancient Greeks testify to honey's widespread use for embalmment. Hindus might hang an urn containing honey and bones from a tree while Egyptians and Babylonians actually placed their corpses in honey jars. The Royal Navy famously pickled Nelson in rum for his last voyage but ancient generals were despatched in amphorae of honey, like the Spartan Agesipolis after fighting in Egypt;[29] Alexander the Great's body was unfortunately hijacked before it reached home.

Herod the Great had his beautiful wife, Marianne, executed but then kept her in a honey tub for seven years 'for he loved her even in death'.[30] In Judaea's dynastic turmoil, his descendant Aristobulus suspiciously drowned in Jericho after taking refuge with Cleopatra, whereafter Mark Antony shipped him back home in honey.

Beeswax was another highly valued but sometimes unsuccessful cladding. Livy[31] tells of the seventh century BC King Numa and his books. In matching lead coffins, the body disappeared but 'watch candles' kept the books fresh. Less plausibly, Pliny tells of seeing an embalmed hippocentaur brought from Egypt by Claudius Caesar.[32] In the Middle Ages high status corpses were coated in wax and honey, which preserved the English King Edward I for at least 400 years.[33]

22 Homer *The Odyssey* Ch 20.

23 See my poem 'Beware Extravagant Hospitality' in Ch 13.

24 *The Danish History*, transl. Oliver Elton (1905), p44.

25 Retailed by Purchas, p165-6.

26 Homer *The Iliad* (1951, transl. Lattimore), Book 19.33.

27 Euripides *Iphigenia in Tauris*.

28 Aeschylus *The Persians*.

29 Diodorus *Historical Library*, 15.93.6.

30 Josephus *Jewish Antiquities*, XVIII.

31 Cited by Purchas, p164.

32 Quoted by Purchas, p165.

33 Jessie Dobson 'Some Eighteenth Century Experiments in Embalming', in *Journal of the History of Medicine and Allied Science*, VIII, October 1953, p431.

Wax images were often made of the eminent dead in Roman times – not so different from our Madame Tussaud's. But what followed is certainly distinctive. After a Roman emperor's cremation with sumptuous exequies, a wax likeness was couched under a cloth of gold on the palace porch, 'pale as if it had been a sick person'. For seven days the effigy was attended with sorrowful countenance by black-clothed senators and their white-gowned wives, before being moved to a grand catafalque in the market for lesser nobles to pay their respects. Physicians had daily:

> … *made shew of feeling the sick parties pulse and always reported that he grew* *worse and worse, until in the end they said he was dead.* [34]

Honey and wax continue into more recent funeral customs. In the Vosges mountains, people might make a candle for their own funeral. A Flemish folk song credits beeswax with driving evil spirits from the corpse but Sardinians refused to escort a bee-stealer's bier with wax tapers. Sometimes honey cakes or other items from the funeral repast might be taken to the owner's hives. In contemporary Russia, mourners will eat around the grave at set intervals and leave honey and other foods there. [35]

The Afterlife

Homer's heroes generally thought Hades a wan place where the dead fill their time with memories of past glories and grand funerals. *The Odyssey* closes with Agamemnon and Achilles reminiscing thus:

> *You were burned in the clothing of the gods, and abundant ointment and sweet* *honey, while many Achaian heroes moved in armour about the pyre…* [36]

Virgil's view is more positive. In *The Aeneid*, while Dido prepares her pyre, priestesses give her poppy seeds and honey drops for her journey to those Happy Isles where golden apples grow. Aeneas himself later finds his father Anchises in the Homes of the Blest, in the river Lethe's beautiful wooded valley where:

> *… like bees in a meadow on a fine summer day settling on flowers of every kind,* *when lilies gleaming white are sprinkled everywhere and all the fields are noisy with* *the hum, the souls of countless tribes and nations were flitting.* [37]

Norse afterlife is distinctly rumbustious with a nanny-goat, Heidrún, keeping the immortal heroes happily drunk on a daily tun (cask) of mead. 'That is a proper wondrous goat for them', says Gangleri. [38]

Bees, perpetual motifs in jewellery, are found among Greek grave goods. One famous treasure is the Minoan gold bee pendant shown in Fig. 14-2. Among treasures recently found in Macedonian Philip II's royal tomb at Aegae is a breathtaking gold diadem, with the insects floating and trembling on fine wires above the foliage (See Fig. 14-3). Sometimes they are explicitly identified with departing souls, as in the collection of gold bees found in a tomb in Peristeria. [39] Those emerging from

34 Purchas, p160.

35 Personal experience.

36 Homer *The Odyssey* (1996, transl. Fagles), Bk.24.66-68.

37 Virgil, *The Aeneid* (transl. WR Jackson-Knight 1956), Book vi, 709ff.

38 Snorri Sturlusson *The Prose Edda*, Gylfaginning chapter, section 39.

39 Marco Giuman (2008) *Melissa: archaeologia delle api e del miele nella Grecia.*

Fig. 14-2 : The Malia bee pendant, 1700-1550 BC, Heraklion
Archaeological Museum, Crete
Courtesy of Ancient History Encyclopedia (www.ancient.eu.com)

Fig. 14-3 : Diadem of Meda, wife of Philip II
© *Archaeological Museum of Vergina - Hellenic Ministry of Culture and Sports – Tap Service*

carcases were seen as the life of the defunct animal,[40] but their sweet and virtuous reputation lent itself to the idea that they were swarms of human souls rising. This belief featured strongly in Orphic and Mithraic cults but Ransome found the same beliefs in Assam, and in Timor regarding warriors' deaths.

Remnants of these ideas carry over into the Christian era. For Celts and Saxons, bees are messengers between worlds and a swarm alighting on a dead tree portends the owner's imminent death. Mouffet accepts the Pythagorean idea that the souls of men and other wise and ingenious creatures pass into bees.[41] In the early twentieth century several such beliefs persisted. For example, in the Swiss Engadine in the late nineteenth century, bees might be seen as souls returning to portend a death.[42] Two Breton beliefs associate them with the process of dying: firstly, Jesus's tears on the cross became sweet and blessed bees; secondly, the soul leaves the corpse as a little fly to prepare for its journey by eating honey from the funeral meal.[43] Ransome heard of alarmed onlookers waking a sleeper as bees flew in and out of his mouth, only to be told resentfully he was on the brink of some paradisal scene.[44] After her death in 2004, various European newspapers depicted the British Queen Mother as a queen bee carried heavenward on apian wings.

A German poet's couplet, despite its sceptical tone, bears clear echoes of ancient Middle Eastern rituals. After appreciating honey's healing powers during life, he concludes:

In that last hour when the soul departs
What help then is honey in the mouth?

TELLING THE BEES

Still common in nineteenth century Europe and the USA, the old customs of telling the bees are maintained by some modern beekeepers, perhaps as quaint relics rather than from conviction. Long ago an anonymous Greek hand wrote:

Naiads and chill cattle-pastures, tell the bees when they come on their springtide way, that old Leucippus perished on a winter night... and no longer is the tending of hives dear to him...[45]

They feel such inconsolable grief at their keeper's death, declares Mouffet, that no 'ringing or tinkling of the brasse pan or any harmony' can delight them.[46]

In a study of growing human sensitivity towards domestic animals, Thomas quotes one Sir William Petty (undated) that the duty to tell implies 'their souls seem… like the souls of men'.[47] A French belief was that they would die or sting if sworn at. Certain language traits further signify their status as 'family'. The Austrian beekeeper, uniquely among stockmen, is known as the *bienen-vater* and greets them in human terms, '*Grüss Gott!*'. Ransome[48] also remarks that the French, too, distinguish bees' deaths with the human verb *mourir* rather than *crever*, as applied to other animals. Germans were unwilling to buy a dead man's hives, fearing his bees would quickly follow. Cook also quotes:

40 A.B. Cook (1895) 'The bee in Greek mythology', *Journal of the Hellenic Studies* 15.1ff.

41 Thomas Mouffet (1658) *Insectorum*, p896.

42 A.B. Cook 'The bee in Greek Mythology' in *Journal of Hellenic Studies* (1895) 15, p1ff.

43 Bertram S. Puckle (1926) *Funeral Customs*.

44 Ransome (1937; 2004 ed.) p223.

45 J.W. Mackail *Select Epigrams from the Greek Anthology*.

46 Mouffet, p901.

47 Keith Thomas (1983) *Man and the Natural World: Changing Attitudes in England 1500-1800*, p98, quoting from p.ii.29 of *The Petty Papers* (1927), edited by the Marquis of Lansdowne.

48 Ransome, pp170, 239.

Little bees, our Master's dead,
Leave me not in my time of need.

French hives were consoled with one of their late master's garments. Said to be silent for six months, hives in the Côtes-du-Nord wore mourning; in Poitou a mourning token would prevent them from attacking their master's ghostly visitations.

Similar customs endured in England. One servant blamed the family for failing to tell the bees that their mistress had died, so the crêpe she put on the hives failed to prevent them absconding. Puckle[49] says the solemn messenger must first knock with the house keys. Dunning[50] writes of an old beekeeper, before his last journey into hospital, struggling round his hives with 'something like affectionate ceremonial', tapping on each in turn. This nineteenth century rhyme accompanied the jingling of keys:

Honey bees, Honey bees, hear what I say!
Your Master J.A. has passed away.
But his wife now begs you will freely stay,
And still gather honey for many a day.
Bonny bees, Bonny bees, hear what I say.[51]

Devon had a contrary custom, turning the hives' backs on a departing bier. The *Argus* reported a disastrous misunderstanding on 13th September 1780. A large squadron of horsemen waited upon a well-to-do Cullompton farmer's hearse. Hearing a sudden cry to turn the hives, a servant set them on their side. Mayhem ensued.

The disturbed bees attacked horses and riders. In vain they galloped off, the bees as precipitately followed, and left their stings as marks of indignation. A general confusion took place, attended with loss of hats, wigs, etc., and the corpse was left unattended; nor was it till after a considerable time that the funeral attendants could be rallied, in order to proceed to the interment...

A century later, the *British Bee Journal* reported that Bradford hives would receive funeral invitations in the same form as family and friends. Betsy Powner wrote to the *Daily Mail* in 1911 about the bees' vigil after her grandmother's death, settling on her bed curtains until 'directly she passed away they all went back through the window, but not to the hive. They all disappeared.'[52]

That tale reminds me of another, where bees were said to follow the hearse to the grave. Indeed we probably underestimate 'dumb beasts'; there are several tales of dogs and even a cat attending to their departed owners. One story tells of a dog finding its own way some distance to its master's grave and being discovered there two days later. My own cat moved from the foot of the bed to sit on the pillow beside my partner's head for her last quarter hour and then immediately moved away, her duty done.

Mark Twain's Huckleberry Finn knows 'the bees would all weaken down and quit work and die' if not told. Also from the USA, Whittier's poem 'Telling the Bees' recounts his beloved's premonition of her own death as, 'draping each hive with a shred of black', she sings:

49 Bertram S Puckle (1926)
Funeral Customs.

50 J.M. Dunning (1945)
The Key to the Hive, p7.

51 Andrew Gough's *Arcadia*
website.

52 L.R. Croft (1990)
Curiosities of Beekeeping, p31.

Stay at home, pretty bees, fly not hence!
Mistress Mary is dead and gone!

But Emily Dickinson is typically restrained:

I have not told my garden yet,
Lest that should conquer me;
I have not quite the strength now
To break it to the bee.

A variation concerns a Sussex woman whose baby died, reportedly because she failed to tell the birth! Kipling's 'The Bee Boy's Song' insists that bees must be told everything or they will desert.

Marriage, birth, or buryin',
News across the seas,
All you're sad or merry in,
You must tell the Bees,
Tell 'em coming in an' out
Where the fanners fan,
'Cause the Bees are just about
As curious as a man!

15 : BEES IN THE BONNET – FOLK CUSTOMS AND CURIOUS ANECDOTES[1]

THE CHRISTIAN YEAR

Bees' seasons govern keepers' timetables but custom often links them to the ritual year. So sixth century Pope Gregory's feast day sees the flowers open on 12th March and a few days later Saint Benedict summons bees to gather nectar. On Passion Sunday, Bohemians rapped on their hives to waken the occupants to work. In England it was wrong to move bees on Good Friday, while Germans thought Easter the right time to burn old hives and medicate sick bees with the ashes. Various countries marked Palm Sunday, Good Friday and Corpus Christi with crosses, wreaths and prayers over the hives; at Easter these were to prevent precious swarms absconding, whilst in France an altar candle was burned to safeguard them. Scottish beekeepers put sugar on the hive step for a midnight Christmas feast.

The insects' own behaviour is sometimes endowed with liturgical significance; for instance, on Christmas morning, Cumberland bees would sing Psalm 100 and those in Monmouthshire apparently hummed with particular beauty.

Precious wax candles, specially blessed, light the procession celebrating Mary's ritual post-natal purification (Candlemas). In many places the crosses from those candles would be thrown into wells to deter satanic influences and witches. Emerging from the Easter Vigil in the small hours, Greeks still take home lighted candles with their eggs. Romanians decorate their dyed eggs with bees.

Martyred by flaying and thus chosen as the patron of tanners, the apostle Bartholomew has a happier connection with the honey harvest; at his Feast the people of Gulval in Cornwall bless their mead and Londoners enjoy honey-coated apples at his Fair.

Many of these hallowed associations have faded over the years and fragrant candles more often ornament romantic or other personal celebrations. Electric fairy lights adorn the Christmas tree but candles in a Yule wreath still light the Christ-child to a welcoming door. A complex tradition survives in modern Macedonia:

It's Christmas Eve — time to set
Wax candles at the door,
Spread honey on the Yule log,
Kiss one another tenderly
And lick the honey from the log
Before it's burned to warm
And light our loving hearth.

1 The stories in this chapter are gleaned from many sources, but I am particularly indebted to Ransome, Fife and Crane.

SAINTS AND BEES

Many of Christendom's legends, miracles and superstitions have obviously earlier sources. The motif of honeyed wisdom occurs in the prophecy of sixth century Saint David's birth; an angel gives thirty years notice by telling his future father he would find a honeycomb on the morrow. In due course it was Modomnoc, one of David's missionaries, who was miraculously helped to introduce bees to Ireland.

> *Setting sail from Pembrokeshire,*
> *Three times the venturer turned back,*
> *Stopped by a settling swarm. Only when*
> *He and all the bees were blessed,*
> *Did they agree to voyage on and*
> *Pour ambrosia on a beeless land.*

As a child, Saint Patrick turned water into healing honey. Later, during a missionary voyage, God saved him and his starving crew via a miraculous delivery of honey and a roast pig. Another story from sixth century Ireland concerns Saint Gobnet, who became patron of beekeepers after her charges' doughty intervention:

> *Between prayers and holy offices,*
> *Gobnet emerged each day to tend her hives.*
> *One morning seeing rustlers among*
> *The cattle in her neighbours' fields,*
> *She deliberately upset her skeps. Loosed*
> *And raging bees stormed the thieves who, all*
> *Forgetful of their ill intent,*
> *Dropped their droving sticks and*
> *Fled with flailing arms.*

When the local chieftain, Herlihy, granted Gobnet's convent land in County Cork's Derrynasaggart mountains, her tiny, gold-jacketed creatures became his gallant soldiers and a skep became a brazen helmet to guard his noble head. To this day, Herlihy's descendants are custodians of Gobnet's Ballyvourney shrine: its modern statue stands on a hive.

BEES AT PRAYER

Several legends tell of bees guarding a consecrated communion wafer. A delightful Lower Rhineland tale recounts how thieves broke in and stole sacred vessels from Altenburg Abbey. A passing swarm, finding the loot under a thorn bush, protectively covered the Host with comb, an event now memorialised by the Bienenkapelle. A French peasant cured his sick bees by taking them the Host; once he dropped the sacred gift and they came out to clothe it in wax.

More elaborate stories tell of bees building a chapel in their hive, complete with windows, door, steeple and bells, where they guarded the Host, sang most sweetly and prayed at the ritual hours. The first is of Saint Medard,[2] instructing his bees to

2 A devoted sixth century French bishop, reputed to be most generous with alms. From Butler and various web pages on saints' lives.

forgive a penitent who had tried to steal them. Another concerns an Irish priest who, distracted by the need to catch a swarm, lost the Sacrament on his way to a sick parishioner; he remained distraught for a year until an angel told him bees had taken it and built a place of wonder containing two priests. Yet another was of a simple woman whose ailing bees made a splendidly productive recovery after she put a Host among them.

Even Charles Butler,[3] that sober Anglican student of bees, recounted such tales, of 'a certain little God-amigti… in a Wafer-cake' being sheltered in a bees' chapel, as testified by one bishop's many miraculous cures with it. He dismisses sceptics; since a blacksmith can make a lock and key so small a fly can operate it, why would these wise insects find a chapel impossible to build? Remnants of these stories travelled to the USA, for instance with a priest putting a fragment of 'The Little God Almighty' in each hive.

In some Slav lands, holy images often protected hives from evil. Variant Slav legends tell of Jesus and Peter asking a woman for a cake from a batch awaiting the oven. One marked by Jesus's finger becomes miraculously large, so she wants to keep it and indents another but Jesus is not fooled. Throwing the bun at him in the ensuing argument, she wounds his forehead. From that wound a beautiful grub emerges to become a bee, which flies away to make candle wax.

Russian churches desperate for candle wax sent Saint Zosima[4] to Egypt to find some bees. The foraging season begins and ends with ritual pleas to their saintly patrons for the bees' safety and good honey.

Such a wealth of legend and folk ritual must be rooted in archaic ideas of bees and the sacred.

Better a Handful of Bees than a Basketful of Flies

This Moroccan proverb sums up the central tenet of many moral fables, contrasting virtuous, fastidious bees with despicable insects. Romanians tell thus of bees' origin: when two children heard that their mother lay dying, the son was too busy at his loom to take notice but the bricklayer daughter rushed home. His mother's curse turned the son into a spider, doomed to sulk in the corner weaving endlessly, uselessly, until it was brushed away and killed. The loving daughter, on the other hand, received a blessing that turned her into a bee, sweet and full of precious gifts.

Moral blame is very occasionally attributed to bees themselves. For instance, a ban on foraging in red clover is because they broke the Sabbath.[5] Other stories chastise them for pride or evil ambitions, as in this Aesop fable: having presented Zeus with finest Hymettan honey, the queen bee is promised her heart's desire and she requests stings for her subjects to enable them to guard their homes. Displeased with this entreaty because he loves mankind, Zeus struggles to find a fair solution – which is that the bees will be armed but on pain of death! Aesop's moral is 'Evil wishes come home to roost'.

Kindred Christian tales resurface. For instance, God asks a grasshopper, spider, ant and honey bee what each is doing. They reply in turn – singing, spinning, working to live, kneading all good things. God admires the ant and relishes the bee's honey but, admonishing her pride, decrees her sting will kill her. In another story pride leads

3 Charles Butler (1609)
The Feminine Monarchie,
Ch 1, p50.

4 See also Ch 4 for a little more about this elusive character.

5 In fact, the flower is the wrong shape for a honeybee but is foraged by bumblebees and others.

bees to demand a silver hive but God says a skep of straw and dung is good enough.

Their unique status is honoured in many ways, not only by telling them significant news and addressing them with human terms. It was deemed sacrilegious to kill one, perhaps because of their long association with souls.[6] It was commonplace that sympathy must exist between bees and their owners. Stolen bees or those housed in stolen straw will not thrive, might even die. They may be bartered but selling them brings bad luck. They will not prosper where there is swearing or a quarrelling family; a Lincolnshire idiom for stirring up disputes or scandal is 'He's set the bea-skep in a buzz'.

Bees and the Law

This special status sometimes went as far as prosecuting the erring insects and not only the stockman whose creatures caused harm. The following is the foundation case in Ireland's mediaeval *Bechbretha:*[7] while entertained at Tara by the seventh century High King Domnall, the Ulster heir Congal lost an eye to a bee sting. To avoid his own son being blinded in retribution, Domnall ordered a lottery to select a whole hive as forfeit. In the German town of Worms the Council in 864 AD ordered a hive's public execution by suffocation after a bee killed a man. Again, in the sixteenth century, a hive was condemned to burning in the public square when a child died of stings.[8]

Perhaps less surprisingly but surely difficult to police, landowners could claim a fixed share of honey from trespassing bees. They could lawfully forage only the area a cow would graze before milking time.

But they could also be grisly allies of the penal authorities. In Constantine's time, Julian the Apostate punished a Christian for pulling down a pagan temple; after scourging and smearing with honey he was confined in a basket of angry bees. In the Spanish West Indies a similar fate met couzenors and counterfeiters; naked, covered in honey and bound, the culprit would be:

> ... *tormented with Flies, Bees and scorching beams of the sun, he might endure punishment, pain and death, due to his lewd and wicked life.*[9]

Topsell says that shows Spain's barbarous colonial policies but his contemporary, Purchas, points out that in Seville, a woman who beats her husband would be paraded on an ass, half-naked and similarly smeared.[10]

Bees might also be used to test innocence, as in the myth of Comates,[11] a Sicilian shepherd who sacrificed one of his master's flock to the Muses so that the rest might prosper. Shutting him up in a hollow tree for a year, the master says 'Let's see if they'll feed you'. Vindicated when the bees build comb around him, he survives the sentence. Among some African tribes, an accused person proves his innocence if his hand escapes unstung from a mass of bees. Commonly in Eastern Europe a girl would show her virginity by walking unharmed through an apiary, while in Slavonia and Bosnia a child would gain a magic skill in handling them if introduced to the apiary with clanging fire tongs and kettle.

Clare Leighton's modern woodcut (See Fig. 15-1) parallels Gilbert White's anecdote overleaf. Maybe the insects recognised the innocence of Selborne's 'idiot boy', as he was seemingly immune to stings. Seizing them bare-handed, he would:

6 See Ch 14.

7 Fergus Kelly and T.M. Charles-Edwards (1983 ed.) *Bechbretha,* Clauses 30-31.

8 F.A. Beach (1975) 'Beasts Before the Bar' in Alan Ternes (ed.) *Ants, Indians, and Little Dinosaurs.*

9 Edward Topsell (1658) *The History of Serpents,* p646.

10 Samuel Purchas (1657) *A Theatre of Politicall Flying Insects,* p161-2.

11 Purchas, p164. Attributed to Theocritus, a Greek poet of the third century BC.

Fig. 15-1 : Clare Leighton's 'The Lunatic Boy'. WA 1992-120
© *Ashmolean Museum, Oxford*

… at once disarm them of their weapons, and suck their bodies for the sake of their honey-bags. He was a very bee-bird, and very injurious to men that kept bees; for he would slide into their bee-gardens, and, sitting down before the stools, would rap with his finger on the hives, and so take the bees as they came out. He has been known to overturn hives for the sake of honey, of which he was passionately fond…[12]

Two strikingly similar stories two centuries apart tell of apt revenge on grasping parsons who demand a tithe of bees in addition to the usual honey portion. When the first beekeeper asserts their number cannot be told, the parson requires 'the tenth swarme, and ye were best bring them to my house'. Two days later, the keeper arrived with a great swarm at the parson's supper table:

'I mary', quoth the parson, 'this is neighbourly done; I pray ye carry them into the garden.' 'Nay, by troth… I will leave them even here.' With that he gave the Hive a knock against the ground, and all the Bees fell out: some stung the parson, some stung the wife, and some his children and family; and out they ran as fast as they could… leaving their meat upon the table. The gent went home, carrying his empty hive with him.[13]

12 Gilbert White (1789)
*The Natural History of
Selborne.* Letter dated
12th December 1775.

13 Quoted in Wm Cotton
(1842) *My Bee Book*, pp102-3,
from a 1593 publication by
Margaret Astley.

The parson's defeat does not end there as the bees decamp into a hedge and are found next day 'by a poore man, who since hath had good profit of them'. That is still not the end, for the parson next takes the beekeeper to law for causing injury. The magistrate accepts that the hive, of course, is the defendant's and it was the bees, brought as bidden, that caused the harm.

In the second case, in Somerset in 1794, the carrier caused similar mayhem by leaving them with the words: 'Measter says, if a's to vind the bees, and be rot to thee, a cant too vind the hives!'[14]

ROBBERIES

Bees could be useful guards against thieves. Virgil's servants hide his valuables among the hives when plundering soldiers raid his farm. Chaucer recounts a chaotic scene after a fox steals the Widow's precious cockerel Chanticleer.

> *Ran cow and calf, and eek the verray hogges,*
> *Sore aferd for berkyng of the dogges,*
> *And schouting of the men and wymmen eek,*
> *Thay ronne that thay thought hir herte breke,*
> *Thay yelleden as feendes doon in helle;*
> *The dokes cryden as men wold him quelle (kill);*
> *The gees for fere flown ouer the trees;*
> *Out of the hyve cam the swarm of bees.*[15]

In sixteenth century war-torn Thuringia, when the Minister of Elende's eloquence failed to deter pillaging peasants, his domestics rout them with a barrage of hives. Alas, we are not told how easily the enraged bees are pacified and recovered. [16]

Law officers could themselves fall foul of angry bees. A commonplace book laconically includes an American frontier newspaper snippet that irresistibly recalls Wilde's famous line, 'To lose one parent, Mr Worthing, may be regarded as a misfortune; to lose both looks like carelessness'.

> *…the United States lost one of their most upright and erudite judges by bees, which stung him to death in a wood while he was going the circuit. About a year afterwards, we read in the same newspaper 'We are afraid we have just lost another judge by bees,' and then followed a somewhat frightful description of the assassination of the American Blackstone by these insects.*[17]

Bees could also thwart the law. Hamble River smugglers around 1900 would hide kegs in empty skeps whose entrances had been smeared with honey. Thus the hives buzzed convincingly and deterred closer examination by visiting Excise men.[18]

It is not unknown for bees themselves to burgle human enterprises, drowning in a brewery and producing strangely coloured honey after raiding a sweet factory. In 1911, the *Daily Mail*[19] reported that, aloof unless disturbed, masses regularly gorged to death on syrups in a Cardiff fizzy pop factory.

14 Report in *The Sun* newspaper, March 1794.

15 From the Nonne's Preste's Tale, in *The Canterbury Tales*.

16 From William Jardine (1840) *The Naturalist's Library*, vol 6, p1955.

17 Frederick Locker (1879) *Patchwork*, p71.

18 Letter to *The Times* in January 1953.

19 *Daily Mail* 11th August 1911.

OF BEARS AND BEES

Bears are fittingly named 'honey eater' in Slavic languages. Everyone nurtured on the Christopher Robin books will remember what troubles befell Winnie-the-Pooh in his insatiable greed. The very first story begins with him under a tree listening to loud buzzing high above.

> *And the only reason for making a buzzing-noise that I know of is because you're a bee... And the only reason for being a bee that I know of is making honey... And the only reason for making honey is so as I can eat it.* [20]

Among other strange tales, Purchas asserts that bears:

> *... much tormented with their eyes, which is a principal cause why they so much desire to eat honey out of the hives, that being prickt and stung... their throat especially receives a phlebotomy, or blood-letting for they have no better remedy to ease their brains and eyes burdened and overcharged with humours.* [21]

20 A.A. Milne (1926)
Winnie-the-Pooh, p4.

21 Purchas, p167.

Whether sore eyes or appetite drive them, bears appear raiding hives in illuminated manuscripts, on a column capital in the Doge's Palace, Venice, even on a marble fireplace in Malton, Yorkshire. Sometimes the bees successfully repel them, as in a

Fig. 15-2 : Bear with bees and hives, from fifteenth century Italian MS
© *The British Library Board. All rights reserved. BL025462*

recent wildlife film showing a bear cub driven from a wrecked hive. After rescuing her cub, the mother managed to grab only one comb before leaving them to gorge on the ruins.

Other misfortunes may befall. Jonstonus[22] reports 'dry hollows' (surely he must mean tree nests?) filled with honey from flower-filled Baltic fields, so deep that bears have drowned in them. Versifying an Aesop Fable[23] in 1668, John Ogilby warns rulers whose imperial appetites outreach their competence. Having been stung by a 'too waspish Bee', Bruine vows to 'flatten twelve waxen cities and seize their treasure' but then

… *Bruine's ugly Visage did not freeze,*
Nor his foul hands want Gloves:
The monstrous Bear you could not see for Bees,
No Bacon Gamon was so stuck with Cloves:
Who Hony loves
Not with sharp Sawce agrees,
Ore–power'd by multitude, and almost slain,
He draws his shatter'd Forces off again;
Then said; I better had endur'd the pain
Of one sharp Sting, than thus to suffer all;
Making a Private Quarrel National.

Fig. 15-3 : Russian bear trap. From Krünitz, JG (1774) *Das Wesenlichste der Bienen-Geschichte und Bienen-Zucht* Berlin: Joachim Paul

In managed forests, beekeepers employed sophisticated measures against raiders. Fig. 15-3 shows a bear, having opened the hive door in the trunk, caught on a counterweighted platform. When shot, the bear will fall onto sharpened stakes. A Norwegian variation was, according to Topsell, to anoint a vacant tree with honey so bees would build there and entrap the marauders. Then, they sawed the tree almost asunder so that, 'when the Beast climbeth it, she falleth down upon piked stakes laid underneath to kill her'.[24]

Bears are not the only chancers. Crane saw Indian beekeepers troubled not only by them but also by flying squirrels and, particularly, monkeys. Acquiring the taste from spillages, these cleverly observed how to open a hive and remove a comb. Next, after carrying their loot up a tree and throwing it down to knock off the bees, they would clamber down and feast unmolested!

Two other examples follow of conflicts between bees and their animal enemies. Fig. 15-4 shows the Biblical hero David attacking a lion that had upset a skep. Perhaps aware of bees' fastidious reputation (though their reported delight in urinous farmyard puddles suggests otherwise), the African *ratal* or honey badger is said to repel them by breaking wind into the hive before robbing it.[25]

22 Johannes Jonstonus (1657) *A History of the Wonderful Things of Nature*, p244.

23 This fable mocked the Persian King Xerxes' hubris. The Greek historian Herodotus described his dismay when his bridge of boats across the Hellespont foundered in a tempest.

24 E. Topsell (1638) *The History of Serpents*, p32.

25 Claire Preston (2006) *Bee*, p119.

Fig. 15-4 : David defeating a lion, from The Hours of Engelbert
of Nassau (c 1489, Flanders) MS. Douce 220, fol. 184r
© The Bodleian Libraries, The University of Oxford

CHARMS AND CURSES

Bee products have been endowed with magical powers. The *Rig-Veda* prescribes combinations of honey and incantations to protect against snake and insect poisons or, mixed with liquorice, as a way to secure a woman's love.

In Germanic lands, an unproductive field would be exorcised with honey.[26] Instructions are precise. Clods cut from each corner were to be dressed with oil, meal and honey, and sprinkled with holy water. Then four masses were held. Before sunset as ploughing began, people recited charms and placed all kinds of meal in the first furrow, at the end of which the farmer should find a jar of honey.

Adonis was addressed with wax votive models of injured body parts. The custom continued with prayers to Christ or patron saints, either for a cure or in thanksgiving. Stories were many and widespread of people measuring the length of the votary, presumably with some fertility issue in mind! A woman facing a difficult childbirth measured her circumference and then set up an appropriately sized candle to Saint Margaret.[27]

Figurines were common in black magic, images being baptised and then tortured in curses to bring about a victim's death. Again, the precedents were ancient – Ovid mentions them. In mediaeval Trier a horrifying backlash followed accusations that Jews used them against Christians. The random forms taken by molten wax in water would reveal the future – a lover, a shipwreck…

26 Hilda M. Ransome (1937)
The Sacred Bee, p163.

27 A.E. Fife (1939) *The
Concept of the Sacredness
of Bees, Honey and Wax in
Christian Popular Tradition*.

One particularly convoluted story concerns a crusader and his faithless wife. Her lover was a necromancer who fixed up a wax image of the husband. A sage told the knight what was happening at home. Immersed in a tub of water with a mirror, the cuckold could see the necromancer firing arrows at his image. Three times the knight ducks and evades the arrow but the last one ricochets and kills the adulterer. The affronted husband then sees his wife burying the corpse under her bed. When he gets home she flatly denies everything but he uncovers the body and has her burnt at the stake.

Bees often featured in prosecutions of witches. Contentiously,[28] Murray argued that benign Witchcraft survived alongside Christianity.[29] This blasphemous 'devil worship' was suppressed by torture and excommunication, and a spate of seventeenth century trials such as Salem[30] exposed the authorities' underlying fear of sexuality. In one instance in 1661 the witch appears as a bee – locating the 'mark of Satan' on Janet Watson of Dalkeith, who had met the devil:

> … *in the likeness of ane prettie boy, who had green cloathes upon him and ane blak hatt upon his head.*

Ransome recounts an eerie story of dark deeds from the Scottish Highlands. Two men rest under a tree during a walk beside the River Spey. While the older one sleeps, his surprised companion sees a bee crawl from the sleeper's mouth. He offers his sword as a bridge for the bee struggling to cross a stream, whereupon it disappears into a cairn on the far side. When the sleeper wakes he recounts dreaming of 'a wee, wee crayterie' emerging from his mouth to tell of treasure in the cairn. The wakeful companion rubbishes the dream so vehemently that a quarrel ensues and the mortally wounded dreamer, seeing his foe collect the treasure, vows the tree will bear witness. Although the murderer becomes rich, the tree shadow clings to him and he remains unhappy, single and friendless until finally he seeks a priest's absolution.

My poem below derives from a ghostly American folk tale concerning an Irish immigrant, healed by honey of a sinister visitation.

> *'Mark Flaherty', he heard as he rode*
> *After sunset — a voice calling him,*
> *And at home in bed the same voice*
> *Keeping him from sleep and someone's weight*
> *Sitting on his chest. Overnight his hair*
> *Turned white. Next evening the voice again,*
> *A man creeping up on him. He sickened,*
> *Became a barely walking bag of bones.*
>
> *A beggar then spoke up, 'Mark Flaherty,*
> *Smother yourself in honey from head to toe;*
> *Essential that you gain it for yourself;*
> *Bees' goodness gathered from countless flowers*
> *Will cure you, brown your hair, colour your cheeks*
> *Afresh.' And so it was — he never heard*
> *That awful voice again.*

28 E.g. Rosemary M. Guiley 'Witchcraft as Goddess Worship', in C. Larrington (ed,1992) *The Feminist Companion to Mythology*, p411-24.

29 Margaret A. Murray (1921) *The Witch Cult in Western Europe*. Her theories underpin twentieth century Wicca practices.

30 Dramatised in Arthur Miller's 1953 classic play, *The Crucible*.

BEE BEARDS

Fig. 15-5 :

Lu Kongjiang, in 2011 bee beard competition in Shaoyang, Hunan Province, China

© ChinaFotoPress/ Getty Images

Bee beards occupy a bizarre byway in apiculture. West Virginian James Johnson once held the record for the largest – weighing ten pounds (about four-and-a-half kilos), made up of around 35,000 bees. The photo of Lu Kongjiang in a Chinese competition shows a beard is no longer enough; even so he did not win! Fifty-nine pounds (almost twenty-seven kilos) clung to the victor. [31]

Perhaps the earliest account dates from 1766; a Father Labat tells of:

> ... *a man who called himself the master of the bees; whether he was their master or no, it is certain they followed him as a flock of sheep does their shepherd, and even closer too: for he was entirely covered with them. His cap, particularly, was so covered, that it perfectly resembled those swarms, which, endeavouring to settle, fix on some branch of a tree. He was bid to take it off and did so; whereupon the bees placed themselves on his shoulder, his head, and his hands, without stinging him, or even those who were near him. He was pressed to tell his secret, but all that could be got out of him was, that he was the master of the bees. They all followed him when he retired: for besides those he carried with him, he had legions which attended him.* [32]

31 In Shaoyang, Hunan Province, July 2011.

32 John Mills (1766) *An Essay on the Management of Bees*, p3.

Two contemporaries revealed the trick. Swammerdam showed that bees would follow a queen tied to a pole. Thomas Wildman, a West Countryman in London seeking his fortune with a combination of business flair and showman skills, performed several times before George III with a swarm hanging from his chin. He tenderly tied a silk thread round a captured queen and gorged her on syrup before placing her where he wanted the distraught bees to settle. In a few minutes they would happily and patiently gather round her. Swarms would follow his horseback gallops round a circus ring. He cautioned inexperienced people not to copy him but these acts did wonders for his book sales.[33]

A modern poet, Pat Winslow, offers a sideways look at beards and other bee lore in the following poem.

THE MAN WHO KEPT BEES IN HIS BEARD

No one could refuse a gift
from the man who smelt of honey.
His fingers strummed and stroked.

He sang to them sometimes,
played them songs on his fiddle
when the evenings were damp and cold.

The symmetry of their lives
absorbed him, their geometric cells,
the way they kept things clean.

He had a love of fur.
When they died, he stitched
their bodies together.

It could take years
to make a coat by candlelight.
A hymn to small things, to industry.

Yellow and brown
were his favourite colours.
His furniture smelt of polish.

He could heal a burn faster
than any doctor. A scar might
seem to vanish almost overnight.

And there were jars of mead
on the floor, bowls of flowers
hanging by his door, garlands of herbs.

33 Thomas Wildman (1768)
Treatise on the Management
of Bees, p107-8.

Fig. 15-6 :

Yet none of this could save him.
When he fell ill, the bees left him.
He grew pale and waxy. His eyes glittered.

Finally, the women came to wash his body
and stitch his shroud. There was talk
about shaving. In the end they left it.

At his funeral the swarm stayed close.
The mourners saw them as they filed in,
heard their plainchant above the lintel.

A triad of notes in the key of G.
Crumhorns in the icy mist.
A festival of humming.[34]

Sinister Bees

Occasionally bees are portrayed as dark forces. The Finnish *Kalevala* contains an interesting example. In the course of his archetypal love quest, we find the hero Väinämöinen in Ilmarinen's forge where Mother Nature's sons, Iron, Water and Fire, quarrel. As the smith tries to temper some steel, Ilmarinen asks a bee to sweeten the bitter Water with honey but she wickedly brings snake and ant venom instead.[35] Now Väinämöinen understands why his iron axe bit him.

The amphora in Fig. 15-6 shows honey hunters beset by a mob of angry bees. Scare stories occasionally emerge about the dangers of Africanised hybrid bees bringing more aggression into milder European strains. Scientists say African venom is no more potent but the bees are more inclined to mass attacks. Whether because an aggressive strain has gained ascendancy or the beekeeper has mistimed opening a hive, in adverse weather or when forage is short, I know to my cost that a previously docile colony may sometimes attack.

Fig. 15-6 :
Attic amphora
depicting a honey
hunt, painted by
Swing Painter,
Basel
*Courtesy of Basel
Antiken Museum*

BEES IN THE BONNET

I watched
Warily after that day
They stormed my veil,
Tangled in my hair,
Stung eyes, lips and ears.

Then, bundled into the closed car,
You slapped and beat me
About my head
Until all were dead.

34 Runner-up in the 2009
Academi Cardiff International
Poetry Competition,
published in *Kissing Bones*
(2012), reproduced with
the poet's kind permission.

35 In Bosley's translation,
a wasp fools him and he
accepts its venom in the
belief it is honey.

Several films[36] and fictions have featured 'horror bees'. Roald Dahl's surreal short story 'Royal Jelly' features a long-awaited baby who fails to thrive. The father decides royal jelly would put the child right. Turning all his hives to producing it and adding generous quantities to the baby's feed, he too eats buckets of the stuff and is becoming a bee. The baby, now guzzling voraciously, has become 'fat and white and comatose, like some gigantic grub' that could weigh five tons in five days.

A Bran Tub of Strange Stories, Fables, Irony and Whimsy

Eighteenth-century Christopher Smart's poetic fables, like Aesop's, were critiques of social superiors. This excerpt from 'The Blockhead and the Beehive' is perhaps a barbed commentary on those who confined him, a religious ecstatic, to a madhouse. From a discreet distance, the bard watches Squire Booby tell the queen bee how to manage her affairs. First he suggests the fairies' coach-maker should yoke her bees to a cart for their harvesting; then he disapproves of the hive's architecture and organisation. Wrathfully she turns on this:

> Impertinent and witless meddler,
> Thou smattering, empty, noisy pedlar!
> By vanity, thou bladder blown
> To be the football of the town.
> Finally,
>   … the portal she unbarr'd,
> Calling the Bees upon their guard,
> And set at once about his ears
> Ten thousand of her grenadiers.
> Some on his lips and palate hung,
> And the offending member stung.

The nineteenth century brings a ponderously improving element to popular works. Richard Frankum's fable[37] tells of a bee falling into evil company. Getting drunk in a field with 'honey sometimes… quite up to his knees', he is then enticed by Mr Wasp to desert his family and go to wicked London, where misfortune overtakes him. He drowns in a bowl of sugar water, and Mr Wasp sneeringly abandons him and goes to eat Bee's family. While so much gorging brings on a fatal apoplexy, it is the bee that carries the moral for us all.

> How often man, like this poor Bee,
> While pleasure's path pursuing,
> Becomes, too late, convinced that he
> Has sought his own undoing.

Fig. 15-7 : Cruikshank's drawing of the wasp watching Frankum's bee drown. Shelfmark 280 c 31, Facing Page XXIII

© *The Bodleian Libraries, The University of Oxford*

George Cruikshank fec.

36 E.g. Theodoros Angelopoulos's 1986 film *The Beekeeper*.

37 Richard Frankum (1832, 2nd ed.1861) *The Bee and the Wasp: a Fable in Verse.*

Another book,[38] seemingly written by a schoolmistressy aunt, mixes modern, scientific information with homily. After poor Harriet's earnest lessons it is a relief to turn to the amusing *Buzz a Buzz*,[39] retailed in doggerel verse by a man who confesses to little knowledge of German. However, the illustrations suggest his spirit is true to the light-hearted original *Schmurrdiburr*. Nevertheless, it too is well laced with moral instruction. Scenes from hive life, including a critique of the drone, are soon followed by tales of the feckless Johnny Dull and the awkward amours and picaresque adventures of the people around him. Pigs upset his hives; he sleeps while they swarm; enraged bees make a hedgehog of his thieving son; and so on. Disturbed during a secret tryst, lovers hide in outsize skeps but redeem themselves by skilfully trapping the thieving intruders. The captives are borne off to prison and the furtive lovers receive a parental blessing. Mayday revels are enjoyed by all, even the bees and their queen.

Figs. 15-8 and 15-9 :
Captured thieves.
Shelfmark 280 i 136,
p. 64

© *The Bodleian Libraries,*
The University of Oxford

38 Anon (1825)
Letters on Entomology.

39 W.C. Cotton (1872)
Buzz a Buzz, from Busch
Schmurrdiburr.

William Combe was a sociable journalist who spent years in the debtors' prison. His doggerel recounts the Reverend Doctor Syntax's picaresque adventures while mourning the death of his acerbic, dull wife. While he is enjoying Lady Bounty's hospitality, a swarm settles in his wig. Ignoring entreaties to stay calm, he falls headlong into her lake but his misfortunes continue:

> *The Doctor said, 'While I have breath,*
> *I'll run and not be stung to death.'*
> *Then off his hat and wig he threw,*
> *And up the terrace steps he flew.*
> *Patrick, with impetuous tread,*
> *Flung the hive tow'rds his Master's head*
> *To save his bald pate from the chace*
> *Of this same flying stinging race.* [40]

TWO BOOKISH ANECDOTES AND TWO TRAVELLERS' TALES

And so, though many other curiosities remain untold, I am nearly done with the imaginative uses of bees; I will close this chapter with just four.

Picturing himself as a 'poore Athenian bee' persecuted by cruel bee-masters, George Wither launches a hundred-page diatribe against publishers whose cheating treatment of authors is like burning a hive for its honey. Despite the King's approval, these rapacious traders have landed him in the debtors' prison and now even want his hymn book labelled blasphemous.

> *Good God! How many dung–botes full of fruitless Volumes do they yearely foist upon his Majesties subjects, by lying Titles, insinuations and disparaging of more profitable Books!* [41]

In contrast, Conan Doyle's fictitious Sherlock Holmes retires in *His Last Bow* to keep bees on the South Downs and write a *magnum opus* on the subject, the whole scheme an innocuous façade for signals data to entrap an enemy spy. Two spin-offs from that consummate detective's new-found interest are Allen Sharp's *The Case of the Dancing Bees* and Laurie King's *The Beekeeper's Apprentice*, in which a bright young girl brings an interesting feminist touch to Holmes's retirement.

Perhaps the strangest fancies of all come next. There is a 3000-year-old Chinese report of a bee ten feet in size that lived in the K'un-lun mountains, with a sting that would kill an elephant! [42] Secondly, Johannes Jonstonus addressed a 'compendium of curious information' to illustrious Baltic nobles. [43] His contents spanned from the Heavens to 'bloodless creatures', which included bees (sic). However, at least two items tested his own credulity. One tale concerned geese, [44] solemnly related to his foreign hosts by fourteenth century traveller Sir John Mandevillle:

> *… of the Bernakes… in our country were trees that bear a fruit that becomes birds flying… And… some of them trowed it were an impossible thing to be.* [45]

This delicious notion perchance would get Jonstonus labelled 'a Lyar' but he hedges his bets. Bees in amber are an even greater mystery; when broken open there is nothing but a concave form and thus they 'appear not really, but only from the imaginative faculty of the Heavens imprinted on it'. Nature has manifold wonders – insects and worms breed in plants – and yet, he cannot believe that birds could grow like pears!

40 William Combe (1869)
Dr Syntax's Three Tours.

41 George Wither (1624)
*The Schollers Purgatory
Discovered in the Stationers
Commonwealth,* p19.

42 From the *Erh Ya*, quoted
by Ransome, p52.

43 Johannes Jonstonus (1657)
*A History of the Wonderful
Things of Nature.*

44 A belief known to the
Venerable Bede and vividly
depicted in a fourteenth
century illumination in the
British Library (Harley MS
4751, Folio 36r).

45 'Bernakes' are barnacle
geese. Sir John Mandeville
Travels, (1900 Macmillan
edition, Ch 29). Quoted
from www.bestiary.ca/beasts/
beast1195.htm

16 : WARRIOR BEES

Armies in battle array were often likened to bees, notoriously belligerent in defence of their colony – but there are other reasons for the comparison. Twelfth century Japanese 'whistling ball arrows' buzzed like bees and soldiers said the same about nineteenth century ballistics. Zulu warriors approaching with massed voices crying '*uthuthu*' seemed to the British to resemble a black, buzzing swarm.

NOT MERELY A METAPHOR

Biblical stories claim that bees were active, though not always dependable, participants in the Chosen People's battles. During their demoralising long march out of Egypt, Israelite leaders insisted on keeping faith with God's assurance of reaching the Promised Land. They may have drawn their enemies into perilous, infested defiles to face being routed by naturally enraged insects or, Neufeld[1] assumes, would already consider bagging a sleepy colony to throw into an enemy's home.

EXODUS

Abandoning sacrifice, we trust our One
True God to lead through stormy waters.

Tramping forty years through split seas,
Desert sand, thorn scrub and rock defiles,
The Israelites courage often fails. Slavery
In Egypt was easier than this. They forget
The burning bush, toy with golden idols,
Doubt their leaders' wisdom and their God.
Preachers have to stir their souls anew,
Remind them of the promise —

Canaan
Flowing with milk and honey, if only
They will stay true. He will rouse
Blinding storms of bees and hornets
From clefts in crags and from the scrub.
They must not doubt His might.
Their enemies will scatter in advance,
Their bowmen gain easy victory.

1 Ed Neufeld, 'Insects as Warfare Agents in the Ancient Near East', in *Orientalia* 49, 1980, p30-57.

2 Deuteronomy 1.44.

But sometimes the enemy had the upper hand. In Psalm 118, enemies 'compassed me about like bees', and elsewhere 'the Amorites, which dwelt in that mountain, came out against you and chased you, as bees do'.[2]

Such vague folk memories would eventually harden into writing, not only in the Bible but in other pastoralists' literature such as *The Iliad*. Homer's countrymen embark for Troy in swarms. The life of the hive resonates still in the city states of Classical Greece; in *The Persians*, Aeschylus also uses the swarm image for the attacking armies.

Equally, bees' self-sacrificial bravery is an example to those under siege; for instance, when the Trojans storm the Greeks' palisade, the defenders refuse to abandon their station but fight like bees for their children's sake.[3] In contrast, Virgil characterises Latium's desperate defenders during Aeneas's siege as bees in disarray when their queen is lost:

> *Trapped in their cave by smoke, as though*
> *A shepherd sought to rob their treasure,*
> *They scurry about their frail house of wax*
> *And hiss with helpless fury in the reek*
> *Until the demented queen hangs herself*
> *And they give themselves up to grief.*[4]

Likewise, in the Second Punic War, after Carthage was captured and Queen Dido threatened with rape, her people fled hither and thither like queenless bees.[5]

Conscripted Bees and Siege Tactics

Bees were often directly used as weapons as we will soon see. Two striking etymological clues: the Greek word *bombos* meaning bee is the root word for bomb and bombard; one type of Chinese explosive grenade was called *Qun Feng Pao*, or Exploding Bee Bomb.

Popul Vuh is the Quiché Mayans' creation myth, passed down orally but accreting later features from turbulent events. The book itself probably dates from the mid-sixteenth century, survived the Conquistadors' ravages but was lost until the nineteenth; nevertheless, the story remained a familiar folk memory. The following poem is my free adaptation of a tribal battle with obvious mythic elements. The first line identifies the four progenitors of the Quiché people, whose god Tohil supports them against a large attacking horde.

> *IN GUATEMALA*
>
> *Jaguar Quitze, Jaguar Night, Mahucutah and True Jaguar*
> *Plot their defence against opposing tribes*
> *And build a palisade around their citadel.*
> *Wooden dummies, crowned with gold and silver,*
> *Armed with shields and bows, stand on the parapet,*
> *Few enough to embolden their invaders!*
>
> *Tohil speaks,* I am here. Do not be afraid,
> *And briefs them. Cramming large gourds*

3 Homer *The Iliad*, Book 12, 169-70.

4 My free adaptation from Virgil *The Aeneid*, Book 12.

5 Ovid *Fasti* Bk 3, 551.

With yellow jackets and wasps
They carry them up the mountain beyond
The citadel. The enemies approach
And gourds are thrown. Insects pour
Like smoke into eyes, noses, mouths,
And sting naked arms and legs

Until they run in frenzy, stumbling
Over outcrops, among entangled trees,
Dropping shields and spears, falling
Down the mountain, routed and oblivious
To the Jaguars' arrows and axe blows,
Even to women beating them with sticks.

Entering an era of military manuals, Aeneas the Tactician instructed precisely how to subvert enemy sappers: deep moats would force the attackers, before they breach the walls, to encounter a mass of burning rubbish or havoc-creating wasps and bees.[6] Devastated by Persia's vicious scythe-wheeled chariots, Roman infantry devised elaborate but in the event problematic counter-measures. In addition to preparing siege towers and mounds against the city of Themiscyra, they:

> *… dug tunnels so large that great subterranean battles could be fought in them.*
> *[But] The inhabitants cut openings into these tunnels from above and thrust bears*
> *and other wild animals and swarms of bees into them against the workers.*[7]

Professionalising his army, Macedonian Philip II organised a corps of artillery engineers who developed siege engines and catapults, no doubt useful on his son Alexander's campaigns. Pottery hives were deliberately left about for wild bees to colonise, then gathered up as missiles. Becoming commonplace projectiles on the battlefield, in sieges or naval warfare across the whole region, furious bees smashing into the enemy's midst could cause utter disarray.

ROUT BY HONEY

Apis mellifera presented a different threat to Xenophon's Greeks, campaigning with the Persian Cyrus against his brother Artaxerxes II, in the fourth century BC. Twice Xenophon's men won tactical victories, only to be waylaid by abundant honey. Left in their path by hostile tribes, this caused severe gastric and psychedelic disturbances,[8] and even paralysis; no one could stand up. At best, a small amount made the soldiers seem exceedingly drunk whilst the greediest seemed near to death.

MARCHING IN COLCHIS

In the Pontic neighbourhood, Xenophon's men
Rout their enemy and watch them flee.
Triumphant shouting peltasts rush on
Followed by heavy infantry

6 Aeneas the Tactician (4th century BC) *How to Survive under Siege*, xxxvii.4.

7 Appian *Roman History*, Bk 12.78.

8 Xenophon *Anabasis* 4.8.20. Trebizond honey is known for its irritant and intoxicant narcotic content, from the blossoms of *Azalea pontica* and *Rhododendron ponticum* – Strabo 12.3.18, and Tozer *History of Ancient Geography*. Wilson (2004) *The Bee*, p212, suggests, however, that the greedy army might simply have fallen upon unripe, fermenting honey.

To quarter in prosperous villages.
Gorging on the unguarded benison
Of astonishing hordes of bees, they fall
In helpless heaps, crazed, seeming moribund

As though dire battle had defeated them.
Days pass in stupor. In time the drug wears off
And they march on to Trapezus, welcome
Heroes among cheering crowds.

Among these savage mountains, Strabo says the Heptacomitae, living on wild animals and nuts, lurk in trees like leopards to attack wayfarers. Bowls of honey caught out Pompey's legion, who were then easily dispatched while stupefied. The Romans considered such underhand tactics unethical – a recurring debate ever since armies abandoned their fists as weapons.

Perhaps it was a Christian refinement to use mead? In 946, Saint Olga of Kiev reputedly avenged a relation by plying 5000 Russians with this at his funeral, possibly fortified with Pontic honey, before massacring them. Similarly, in 1439, Russians left casks in an abandoned camp, only to return when 10,000 pursuing Tatars had drunk themselves into a daze.[9]

More Ballistic Hives

These were thrown so frequently in Roman warfare that Ambrose[10] suggests the late empire ran short of bees. Their rebellious colonies seemingly did not suffer the same handicap. After highly successful Mesopotamian campaigns during the Parthian Wars, Septimius Severus suffered a setback when his army reached 'Arabia Felix, so-called because of its aromatic herbs'.[11] The city was actually Hatra and the situation less than 'felix', as furious insects defeated the aggressors.

BESIEGING HATRA

Snug in their stronghold on a sudden ridge,
Ringed by enormous walls, the citizens
Of Hatra smirked as the Roman army
Stood outside, throwing all its weight and skill
To no avail against their citadel.
Every siege technique blunted, each engine
Failed. The soldiers, accustomed to success,
Sweltered under an unforgiving sun
While arrow storms riddled their tortoise shield.
Worse still, clay bombs of noxious flies stung eyes
And every exposed inch. Lethal disease
Slew many more than wounds. They slunk away.

9 J.L. and C.G. Gould (1988) *The Honey Bee*, pp. 2-3.

10 John T. Ambrose 'Insects in Warfare' in *Army* (1974) p33-8.

11 Herodian *Roman History*, 3.9.3-7. It seems Herodian wrongly named the place; Arabia Felix is modern day Yemen and Hatra is in northern Iraq.

Though temporarily defeated by six legs, two legs eventually won the war and Septimius bestowed upon himself the title Parthicus Maximus. Later, Dacians hurling skeps on the imperial frontier (in present day Romania) proved another hurdle. Shedding their earlier ethical inhibitions, when the Romans developed catapults, bee-filled amphorae were among their ammunition.[12]

MEDIAEVAL AND LATER SIEGE WARFARE

The same tactics continued throughout Europe and the Middle East in the defence of mediaeval towns. For instance, in 908, Chester repelled Norse miners by hurling hives into their tunnels and onto their protective canopies. The Archbishop of Tyre wrote of a typical battle for the Syrian city of Maarat al-Numen during the First Crusade:

> The Muslim inhabitants tried desperately to drive the Christian enemy from the fortifications, when they fired beehives at them from their engines of war; they made use of stones, beehives swarming with bees, fire and even quicklime. [13]

This Christian victory was relatively short-lived but Richard the Lionheart captured Acre, thereafter known as the city 'taken by bees'. A thirteenth century poem claims his 200 well-victualled, well-armed ships included, surely implausibly, thirteen shiploads of hives:

> The Saracens then armed all
> And ran anon unto the wall
> In white sheets they wrap themselves
> For the biting of the bees…
> [until]
> Hid in a deep cellar
> That none of them might come near
> They saw King Richard was full fell
> When his bees bit so well.

So routine was all this that people built bee boles ready to hand in castle walls!

In history's endless arms race, the fourteenth century saw a significant advance on existing catapults. A multi-armed wheel was designed to throw a stream of skeps in rapid succession. Scattering cargoes of furious bees, the effect may be imagined as a combination of revolver and machine gun.

When Turkish Amurath II besieged the Hungarian defenders of Alba Greca (Belgrade) in 1439, he was confronted by many hives stationed among the ruins. His Janissaries, 'although the bravest soldiers in the Ottoman Empire, durst not encounter this formidable line of defence'.[14] Africa was not exempt; in Mauritania the citizens of Tornli threw burning hives down onto Lupus Barriga's besieging Spaniards.[15]

During the Thirty Years War, a Swedish army besieged the German town of Kissingen. After thirteen desperate days, the mayor wanted to throw open the gates but the stalwart burger Peter Heil rallied his fellow citizens. Gathering up 200 hives,

12 Conrad Bérubé (2000) *War and Bees: Military Applications of Apiculture.*

13 Ray Jones (2008) *Bee-Sieged: Bees in Warfare,* p55.

14 W. Jardine (1840) *The Naturalist's Library,* Vol 6, *Entomology, Bees,* p194-5.

15 Quoted by Samuel Purchas (1657) *A Theatre of Politicall Flying Insects,* p161.

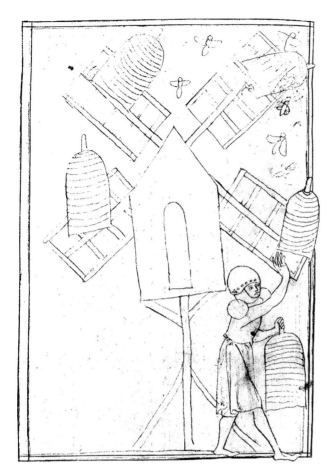

Fig. 16-1 : Cartoon presumably prepared for illumination, from a manuscript by Walter de Milemete (1326), advising Edward III on the arts of kingship

© *The Governing Body of Christ Church College, Oxford*

Fig. 16-2 : Peter Heil, saviour of Kissingen, armed with hives. Sculptor Imre Varga

© *Stadtarchiv Bad Kissingen, Photo Traugott May*

they threw every one over the walls to devastating effect. Although the soldiers were adequately protected, their panicking horses created a rout. An annual summer festival still remembers Peter Heil.

MODERN TIMES

Long after gunpowder was used to breach city walls, bees could still play a decisive role, by accident or design. Anecdotes turn up in diverse places. Discussing the hotter temperament of bees in hotter climates, the Corsican student of bees, Abbé della Rocca, describes forgotten tactics deployed in the early eighteenth century:

> *A small privateer with 40 or 50 men, having on board some hives made of earthenware full of bees, was pursued by a Turkish galley manned by 500 seamen and soldiers. As soon as the latter came alongside, the crew of the privateer mounted the rigging with their hives, and hurled them down upon the deck of the galley. The Turks, astonished at this novel mode of warfare and unable to defend themselves from the stings of the enraged bees, became so terrified that they thought of nothing but how to escape their fury; while the crew of the small vessel, defended by masks and gloves, flew upon their enemies with sword in hand, and captured the vessel almost without resistance.* [16]

16 Jardine, p194.

Tammy Horn[17] includes several incidents from the USA's founding wars. George Washington credited Quaker Charity Crabtree and her bees with 'saving the nation' during the War of Independence. As she passed by with her skep, a dying American soldier accosted her and gave her his horse to warn his General of the approaching British. The enemy being almost upon her, she flung her bees in their path and the resulting panic allowed her to rush onward with the vital message.

Bees also unwittingly played a crucial and unwelcome role in several battles of the American Civil War. On the costliest day of the war, a soldier wrote of his platoon's sheltering farmyard made impossible by swarming thousands amongst the bullets and whizzing shells.

ROULETTE'S YARD: 17th SEPTEMBER, 1862

In grey and blue, casualties strewed
The soil beside the Creek of Antietam.
Artillery scythed The Cornfield packed
With waiting troops. The day ground on
In swirling smoke and stink of powder
While cannon thunder rolled around the woods
And clouds. Nobody could know the end.

Huddled in a bee yard, a company
Of Pennsylvania Volunteers took a break,
Lolling behind a wall, digging in knapsacks
For a bite to eat, lighting their pipes.
They'd fought through mid-day heat.
Now they watched the bees' stolid business,
Tiny bodies loaded with peace

Until a Confederate ball ripped the hives.
Furious hordes besieged the resting men
Who rushed about, blindly beating their faces
And naked arms, redoubling the bees' attacks
And spreading mayhem wide throughout
The Union side. The insects' victory
Almost gave the day to Jackson.

Another time, storms overturned hives and caused such havoc on both sides that the battle had to be adjourned. In Georgia, hives trip-wired by a local woman several times outwitted hungry Union marauders.

On a more homely note, both armies raided hives for a welcome addition to their poor diet. A letter home from Henry Schafer of Illinois describes 'a jolly time in camp' as some of the boys brought in honey molasses [sic] and plentiful livestock.

17 Tammy Horn (2005) *Bees in America: How the Honey Bee Shaped a Nation.*

Cultural Conflicts and Propaganda

David Edwards raises an interesting discussion of different world views, both inevitably propagandist, relating to nineteenth century conflicts in Afghanistan. A British expedition sought a 'fanatic mullah' stirring incomprehensible *jihad* in the Swat Valley. Though he eluded them, the British claimed total success and that, withdrawing with hostages, they brought the villagers' curses down on this 'Mad Fakir' for all the destruction. The Afghans gave a very different account – swarms of bees overwhelmed the British, who told their commander they:

> … were faced with this kind of plague and disaster that all of our bodies, from poison, how painful, and we couldn't reach there. God sent bees against us.[18]

A similar mismatch of accounts occurred in another encounter, when the British blamed bad weather for their defeat while the Afghans thanked God for sending a storm of bees.

Insects Versus the Monstrous Machinery

Despite the First World War's industrialised slaughter, bees featured in fringe campaigns. In *Duel for Kilimanjaro*, Leonard Mosley describes Von Lettow-Warbeck's brilliant guerrilla campaign against the British in Tanganyika.[19] German ambushes drew ill-prepared opponents into forests where wild bees, disturbed by the firing, on occasion routed British and Indian forces. *The Times* criticised such devilish devices but the German commander said bees had disabled several of his emplacements too. In correspondence columns as late as 1953, old soldiers were still discussing whether the nests had been deliberately tripped. Wild bees quite commonly routed both sides, after battles or fires started by careless cigarettes roused them, but one writer vividly remembered trip-wired hives causing mayhem among British columns in Cameroon forests and others said this was an immemorial West African tactic. One ex-colonial officer in Nigeria painfully recalled these Parthian arrows vicariously shot at the 'protecting power'.

A European incident in the same war involved trapped Belgian defenders with admirable phlegm hurling bees at their German attackers. Surely these were not substantial modern hives or puny single frames or historic clay bombs? Perhaps they were skeps. In contrast, this incident in an unknown location was obviously arranged with much forethought:

> AMBUSH
>
> Well guarded by a company of infantry,
> A baggage train laden with socks and food,
> Guns and helmets for beleaguered troops,
> Climbs through a narrow defile. Brushwood
> Blocks their track. Soldiers rush
> To clear the way, tripping a wire

18 David B. Edwards, 'Mad Mullahs and Englishmen: Discourse in the Colonial Encounter' in *Comparative Studies in Society and History* (1999) 31.4, p 649-70.

19 Leonard Mosley (1963) *Duel for Kilimanjaro: The East African Campaign 1914-18*. In defiance of the German ambassador, the Colonel sustained a guerrilla campaign throughout the war, which kept a sizeable British army busy far from the European arena. Some sources name him Von Lettow-Vorbeck.

To a regiment of hives requisitioned
From villages all around and hidden
Among rocks and thickets. Every man
And beast lies dead, and all supplies lost.[20]

World War Two saw Ethiopians dropping hives into the Italians' open tanks, causing terrorised crews to crash into precipitous ravines. All parties in the Vietnam War were mindful of bees as allies or enemies. In *The Tunnels of Cu Chi*, Chi Nguyet is quoted:

> *So we set up some of these hives in the trees alongside the road leading from the ARVN post to our village. We covered them over with sticky paper...*[21]

Stumbling upon this fragile skin, an enemy patrol would enrage the bees and then, scattering, fall into spiked traps and flee, dragging away their wounded. While US forces flew overhead with Agent Orange and napalm bombs, the Viet Cong arranged small heaps of dirt, to delude the American infantry into seeking freshly dug traps. Hidden hives then raised a terrible commotion, the insects attacking mercilessly 'and in no time at all thirty troops were out of action'. Encouraged by this, the Viet Cong started to breed bees specially and claimed to have trained them to recognise US uniforms, which seems implausible! The finer shades of olive drab have not featured among those distinctive colours bees can learn to associate with rewards. Perhaps they distinguished the body odours of men with such different lifestyles. The US military is said to have considered spraying the Viet Cong with bee alarm pheromones to make the creatures turn against their own side.

Fig. 16-3 : Bees being trained as mine detectors
© *Randy Montoya, Sandia National Laboratories, USA*

20 My verse derives from an anecdote recounted by E. Crane (2003) 'Retrospect: use of bees in warfare' in *Bee World* 84.2, pp94-7.

21 Tom Mangold, and John Penycate (1986) *The Tunnels of Cu Chi*, p125-6.

EVENTUALLY PEACE RETURNS

In the 1990s, US defence scientists[22] were researching more benign possibilities. They hoped to train bees via their excellent sense of smell to associate trace explosives with nectar sources and thus use these 'flying dust mops' to locate the disastrous legacy of unmapped landmines that maim and kill many thousands annually around the globe. If suspect foraging areas could be identified, then large tracts of uncontaminated land could be restored to human use. But varied weather and terrain present multiple problems and that programme stopped several years ago. However, Zagreb University is still researching among the 250,000 mines littered around Croatia by the Yugoslav Wars; brave mine clearers and innocent peasants continue to lose their limbs or lives to these buried evils.

Fig. 16-4 : 'After the war comes peace'. From Andreas Alciato (1549) Emblems Courtesy of CAMENA

The implements of war often find new uses. Herodotus was perhaps the first to write of bees colonising battle's debris, seeing combs in a skull left on a spike in Cyprus.[23] A free translation of the verse accompanying the Spanish helmet reads: 'When wars are over and knights return from endless noise to their quiet shires, their helmets stained with enemy blood find rest in sheltered corners, to provide questing swarms with ready hives and delicious combs of honey for worthy veterans who fought just wars in the cause of peace.'

This extract from Whittier's American Civil War poem offers a final poignant note.

THE HIVE AT GETTYSBURG

… A stained and shattered drum
Is now the hive where, on their flowery rounds,
The wild bees go and come.

Unchallenged by a ghostly sentinel,
They wander wide and far,
Along green hillsides, sown with shot and shell,
Through vales once choked with war.
The low reveille of the battle-drum
Disturbs no morning prayer;
With deeper peace in summer noons their hum
Fills all the drowsy air…

22 *Journal of Mine Detection VII/3, 380-9* (2003); *Journal of ERW and Mine Detection 13.1.* (Summer 2009). Recent references to the topic on the web constantly refer back to the University of Montana's work up to 2003-04.

23 Purchas, p163.

17 : FROM LANTHORN PANE TO ELECTRON MICROSCOPE

CURIOUS AND NOT SO CURIOUS BASICS OF ANATOMY AND PHYSIOLOGY

Like other insects, bees metamorphose from egg to larva to pupa and finally imago. The queen takes sixteen days to mature, crucially receiving more royal jelly than her sister larvae to become a fertile adult. Once hatched, she is soon eager to mate and, after one or two nuptial flights, during the active months lays up to 3000 eggs a day for at least three years. Registering the size of the cells with her forelegs, she can regulate laying fertilised eggs in worker cells and her parthenogenetic drone eggs in their few hundred larger chambers.

Workers take nine days longer to reach adult form, and having received only three days of royal jelly are normally infertile. Occasionally they lay eggs if the queen fails but the progeny will almost all be drones. Drones are generally the product of unfertilised eggs and take as long as a worker to hatch, then another fortnight to become capable of mating. But thousands of years passed before bees' complicated sex life was understood.

Adult workers weigh approximately 112 milligrams and can carry their own weight in cargo. Their six legs have specialised tasks: pollen baskets on the back pair, and on the others combs for grooming and ladder hooks used when swarming or building. Abdominal glands exude building wax. As stinging is suicidal, a bee only does so when under threat but then her pheromones rouse the rest of the colony to fury. The last fortnight of her six weeks of adult life is filled with foraging, during which she flies around 500 miles at speeds up to 22 miles per hour, faster than any comparably sized insect. The whole colony will fly around 280,000 miles per day! A nucleus of the colony survives winter by fattening up in autumn and then forming a minimally active self-warming ball within the hive.

The modern Butler[1] summed up the mid-twentieth century understanding of bee senses. The food odour currently in the hive influences their flower preferences and they can be trained to associate flavours. Turning up when jam is on the table indicates a good sense of timing! Despite the age-old insistence on their fastidiousness, they positively prefer cow-dung water to the clean, distilled type.

Each antenna carries 12,000 sensory hairs and 5000 pores, enabling a sense of relevant smells a hundred times more acute than ours. Connected to the hemispheres of an exceptionally large brain, these antennae provide stereo directional guidance among other functions. Feeling vibrations, pressure and air movement, body hairs give positional information crucial to their dance. Sensors underfoot may also 'hear' vibrations. Feet and antennae are particularly sensitive to different surfaces, important for navigating in the hive's darkness.

Each of the worker's eyes is a compound of 5000 units, receptive to a limited colour range and more attuned to movement than shape. In addition, their foreheads carry three simple, unfocused ocelli which react to light levels.

1 C.G. Butler (1954)
The World of the Honeybee.

Recent research has further identified the importance of pheromones: they act like a welcome mat for homecomers, and include a queen substance related to swarming.

Unravelling the Facts, a Work of Many Centuries

Ancient ideas were continually recycled but knowledge slowly accumulated. For instance, Seneca was the first to think bees do not merely gather honey but contribute to its formation.

Pre-Linnaean categories strike modern minds as capricious. Pliny[2] classes bees with dolphins, because neither tame nor wild. Strangely disregarding gnats, a mediaeval Bestiary classifies bees as the smallest of birds. Bartolomaeus Anglicus is perhaps exceptional in the mediaeval period by claiming to work from first-hand observation, although he repeats Pliny's dubious assertion that:

> *... if the night falleth upon them in their journey, then they lie upright to defend their wings from rain, and from dew, that they may in the morrow tide fly the more swifter to their work with their wings dry and able to fly.*[3]

Renaissance minds gradually shift from philosophical and theological reasoning to practical observation and experiment. Two empirical forerunners are the German, Nickel Jakob, in 1568 and a few years later the Spaniard, Luis Mendez; Jakob notes the queen grub's special nurture and Mendez reports that she is the colony's sole mother. In England, Charles Butler[4] brings a remarkable observational rigour to the whole field. He complains about untested notions constantly repeated, though risible to 'every silly woman' with practical experience. Nevertheless, literary flourishes and sometimes fanciful theories adorn his own work.

He describes the hive's different occupants and comes to the shocking conclusion that the Rex is female.[5] She is 'fair and stately... in shape and colour', her back brighter brown, her belly 'deeper than richest gold'; 'golden bars' divide her body whereas workers are marked by silver bars.[6] Bigger and longer than the others but not as broad as a drone, she has shorter fangs and tongue than a worker. 'If some curious chirurgian would make an anatomi', he would find the male parts of the drone. The workers' size 'resembles a cloak rather than a gown'. Foraging with fangs and tongue while four feet stand fast, the other two pack their thighs and groom. Seeming to have no brain, workers have senses 'which their subtil and activ spirits doo excite and quicken... for their curious art and singular virtus'. An excellent sense of smell quickly leads to distant forage. Scent also identifies stranger bees and culls out drone cells. But can they taste? Aristotle doubted their hearing but why then would there be swarm and 'battle musik'? Their 'horns' provide an acute sense of touch; a general noise is excited by any touch in the vicinity of the hive. Instead of eyes, transparent cheeks 'like lanthorn' seem their weakest sense:

> *... for returning loaded they seek the door as if in the dark; fly hither and thither, rub and wipe their glazen eyes that they might better discern their way...*[7]

2 Pliny *The Natural History*.

3 Bartolomaeus Anglicus *On the Properties of Things*.

4 Charles Butler (1609) *The Feminine Monarchie*, Ch 1.

5 If Butler knew of Mendez's finding, England at large certainly did not.

6 Was this true of native English bees at the time or was it one of Butler's several rank-conscious distinctions? Modern worker bees are more banded in brown-black than the yellower queen.

7 Butler (1609), Ch1, para 43.

Butler believes a healthy bee lives a year; others suggesting seven years wrongly base their estimate on a hive's longevity, which is a matter of fortune and no measure. A sick colony may be cured by warmth and food but winter may well finish them off.

Moffett's *Insectorum* (1634) had an extraordinarily chequered history, being inherited by an orphan whose overseers could not afford to publish it and no money-grubbing printer would accept it. And so it lay 'for some years cast aside in the dust among Worms and Moths' until it fell into Edward Topsell's hands. After extensively plagiarising its bee content, Topsell then annexed it to his own 1658 *History of Serpents*.

Fig. 17-1 : Moffett's title page. Shelfmark RR x. 63, Title Page

© The Bodleian Libraries, The University of Oxford

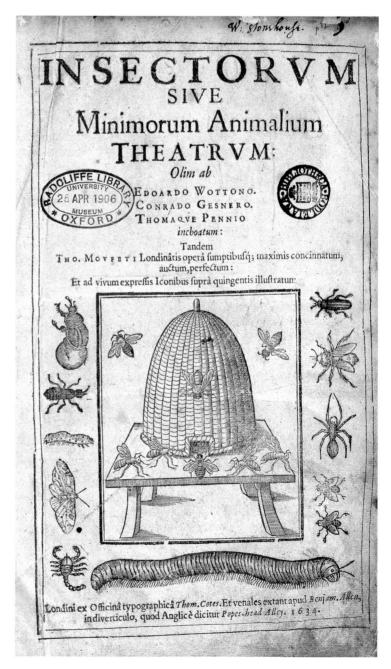

Besides familiar ideas, both distinguish between bees 'domestical and tame much delighted with familiar friendship, custom and company of men' and others:

> ... altogethere wilde, uplandish, agrestial [which] no wise brook or endure them, but rather keep their trade of Hony-making in old trees, caves, holes and in the ... rubbish of old walls and houses.[8]

Sovereign among 'venomous insects (or cut-wasted creatures)' it feeds mankind whereas others serve only to provide medicine, delight eye and ear or ornament the body.

Starting with 'very thin and trembling wings', by age two to three they are 'very trim, bright-shining and in very good plight and liking, of the colour of Oyl' but in old age they become 'hard, thin and lean scrags' – but wise! The authors describe garden bees living among abounding sweet-scented plants as 'great, soft and fat-bellied' [bumblebees?] whereas town and village honey-gatherers are 'lesser in proportion of body, rough and more unpleasant in handling'.

'Flies as Big as Lambs': The Wonders of the Microscope

Three hundred years lay between the thirteenth century's first primitive spectacles and the Galilean telescope. This was then accidentally discovered to magnify objects close at hand, albeit clumsily at three arms' length! Now Galileo exclaims, 'With this tube I have seen flies which look as big as lambs'. Suddenly the horizons of the small expand as marvellously as the cosmic. As ever, bees attract a great deal of interest.

Jan Swammerdam[9] combines scrupulous scientific observation with unusual religious fervour. Long years spent refining the techniques of staining and dissection and poring over primitive Leeuwenhoek microscopes result in superb engravings, still valued today.

Comparing and contrasting bees with other animals, he discovers many details of their anatomy. Air pipes near the wings create their hum; there are unexplained nerve fibres 'packed in the manner of timbers on ships from Amsterdam'. He cannot find organs of hearing or smell, but surely God would have given them those? He fathoms their sexual anatomy but not their reproductive habits. Although only able to speculate about bees' life span, he observes that their worms are incubated simply by the heat of the bees' motion. When fully grown, they spin a cocoon within the cell and develop their parts by slow accretion, not fictitious metamorphosis.[10] He is enrapt by the very tender nymph stage, the eyes changing 'from a limpid or clear white, becoming of a beautiful but somewhat pale purple'.[11]

When ready to emerge, air and blood inflate their folded wings and then the creatures bite their way out. Among miscellaneous details, he reports how venom tastes increasingly acrid, causing salivation and affecting tongue and jaws like highly rectified spirit. Larvae leave 'a very offensive or nauceous rancour like that of rusty bacon in the mouth'.

Pressing one eye against his tiny bead microscope, he drew a remarkable study of the multifaceted form of a bee's eye. After twelve years, this endlessly inquisitive, true scientist regretfully confesses exhaustion but, before laying down his pen, notes

8 Edward Topsell (1658) *The History of Four-footed Beasts and Serpents*, p638.

9 Jan Swammerdam (1758) *The Book of Nature.*

10 Evidently the modern usage of 'metamorphosis' was then unknown; he is rejecting the Ovidian notion of magical transformation. It was only in the nineteenth century that a full understanding of embryology and metamorphosis was achieved.

11 Swammerdam, p172.

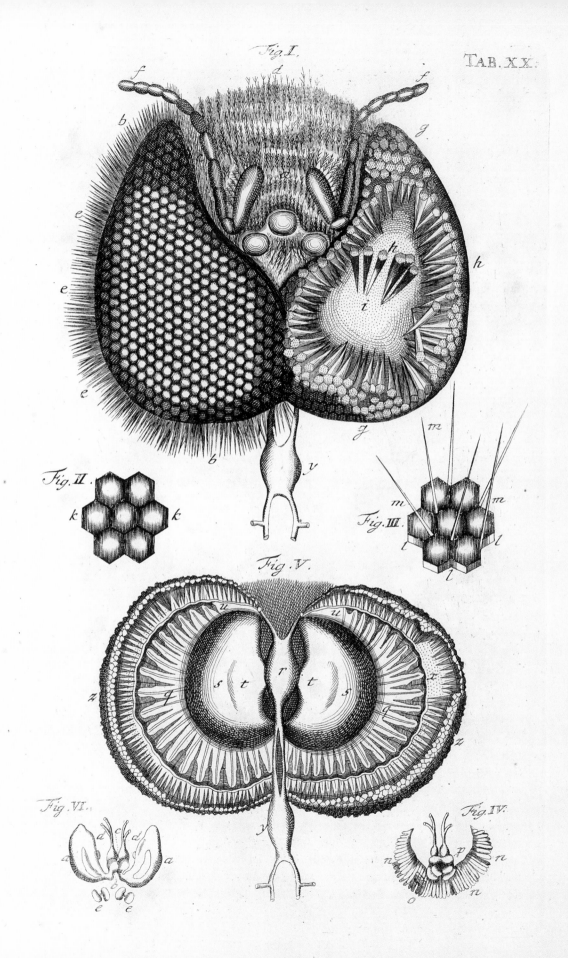

Fig. I.

TAB. XX.

Fig. II.

Fig. III.

Fig. V.

Fig. VI.

Fig. IV.

that 'some subtile and sagacious geniuses', such as the illustrious Dr Hooke, believe insect eyes are 'only a congeries of innumerable little eyes'.

(Left) **Fig. 17-2** : Swammerdam's drawings of a bee's eye. Shelfmark Douce S subt. 48, Tab XX
© *The Bodleian Libraries, The University of Oxford*

THE NAKED EYE AGAIN

Samuel Purchas[12] is another close observer, though apparently without a microscope, affirming some truths but robustly refuting ancient fables. Aristotle was wrong to think there were forty types of honeybee and that their eyes moved. He was equally mistaken in claiming their beauty varies, like that of ladies, with idleness! Not so; it is age and work that scars and makes them bald. They do not carry stones in rough weather but do adjust to wind, sailing on it or flying in the lee of a hedge. In battle, fangs are far more dangerous than stings, which are mainly for hot-blooded creatures and offer a merely Cadmian victory.[13] They do not dislike red, get 'inebriated with sweet oyntments' or offended by stink; his hives prospered for years beside a dunghill.

No white diadem adorns the 'fair and stately' queen but she has a 'lofty pace and countenance'. She is not equipped for any work but laying. That a hive has a king and queen, or two harmonious queens, in a palace fenced and surrounded by a royal guard – all fables.

Evidently bees have five senses, though not all organs are visible. He endorses Butler's view – they have an acute and inquisitive response to touch and are long-sighted, able to navigate several miles but stumbling at the hive door. There they 'take much pains rubbing their horny eyes'.[14] Sound signals clearly indicate they have hearing. Their prudence is not moral but instinctual. They have memory and a learning faculty but no imagination.

Concluding with a 'dish or two of sweetmeats' for judicious readers, Purchas is at odds with most past scholars who deny insects have lungs, nostrils or blood and therefore need no cooling; being invisibly small does not mean they do not exist. You can see them pant on the step and a feather will vibrate at a sleeping hive's door. Debate continues, but he thinks their buzz is their breathing, not the sound of wings. They do sleep, through winter and at night but not when abroad. The truth is that a bee lives not more than a year and a quarter!

In the same mid-century, Moses Rusden's[15] windowed hives further the new observational approach. He was the first to experiment with deliberately hiding the queen and watching the effect. He notices how they orientate themselves to a new hive:

> *… a bee coming out into the sunshine wipes his eyes with his foreleggs, then takes his flight not above three inches from the mouth of the hive, and there flyes three times round; then he flyes further off and flyes thrice round again; and afterwards a pretty distance off and fetches a large circumference three or four times, all this while taking notice, first, of the mouth of the hive… then of the places about it, and at last of the places about the house and garden…*[16]

12 Samuel Purchas (1657) *A Theatre of Politicall Flying Insects*, Ch 2.

13 A reference to the Greek myth of mutual slaughter of the warriors who sprang from the dragon's teeth sown by Cadmus.

14 Purchas, p22.

15 Moses Rusden (1685) *A Full Discovery of Bees*, p8.

16 Rusden, p93.

Like Purchas, he considers them 'not altogether tame nor absolutely wild… indocible, for most they do is by instinct'. But he still writes of a king, generals and soldiers. The king is like a 'most stately Buck Gray hound'; despite knowing Butler's work, he cannot admit a feminine monarchy. The 'common Honey-Bee' is like 'a fierce little bulldog' and lives a year, darkening with age and wearing its wings ragged. The male is like 'a great Mastiff-Dog' but without stings or fangs, so easily killed. Drones warm the brood while workers are out.

Having no visible organs or brain, yet bees have 'senses both outward and inward', i.e. physical and moral. Sight is evident from their flights around a new hive. Subtle touch enables them to navigate in the hive and also to distinguish between a dead thing and an intruding hand to be stung. Their sense of smell is acute, alert to honeydew, roused to mob violence on behalf of a single dead companion, able to distinguish friend from foe. They must have taste because they love luscious things. They respond to the king bee's signal notes and sing joyfully when they locate a lost leader. People bang pans to stay a swarm because it drowns the leader's calls.

A few years later, Joseph Warder, too, assumes a maximum lifespan of a year, but sees too many hives prematurely lost or destroyed because of vastly divergent opinions about bees' longevity. He has no doubt about the monarch's sex and rhapsodises that her beauty so far surpasses that of her subjects 'as the finest Horse that ever ran on Banstead Downs doth the common Forrester'.[17]

Priding himself on being a practical beekeeper, Levett laconically dismisses the question of bees' lifespan: 'I would advise no man to make triall of this matter, except hee like better of curiosity than profit.'[18]

Robert Hooke is an enthusiastic English advocate of the microscope; preferring inferior compound lens instruments to Leeuwenhoek's difficult instrument, he convinces sceptical colleagues in the Royal Society. His *Micrographia* (1665) is another stunning testament to the skill and patience of scientific illustrators, a learning tool that microphotography surely cannot replace.

For a time microscopes were an exciting novelty but years pass before people learn how to interpret this uncharted world or prepare specimens well enough to read accurately. Interest declines and by 1692 Hooke is regretting the 'want of inquisitive Genius of the present Age'. However, it is a great age of instrument craftsmen and enthusiasm revives with the rapid development of pocket-sized compound microscopes.

Henry Baker,[19] a naturalist and poet whose varied life included bookselling and teaching deaf and dumb children, wrote two books about the instruments. He wrote to his Fellows of the Royal Society in 1740:

> *Gentlemen, A desire to excite in Mankind a general desire of searching into the Wonders of NATURE, will, I persuade myself, be accepted favourably by you… Nothing is really needed but good Glasses, good Eyes, a little Practice and a common Understanding, to distinguish what is seen…*

Insect eyes amaze him. Many thousands, 'so perfectly smooth and polished, like so many mirrors'; 'how delicate, exquisitely delicate' the retinal filaments must be compared with those of whales or elephants. Their antennae are so beautiful – branched, feathered, tufted like dandelion seeds – for near navigation, where their eyes cannot see.

17 Joseph Warder (1712)
The True Amazons or The Monarchy of Bees, p65.

18 John Levett (1634)
The Ordering of Bees, or the True History of Managing Them, p46.

19 Henry Baker (1742, 1785)
Of Microscopes and the Discoveries Made Thereby.

'Inimitable Embellishments and Embroideries'

Using only his naked eye, Thorley[20] is another careful observer, even proving the queen's sex by seeing her lay eggs on his hand. Eggs in open brood cells become grubs in a bath of watered bee-bread (pollen). Finally, undisturbed in a sealed cell, there occurs the marvellous metamorphosis. Did the grub die? It is so utterly changed, as though in resurrection. He is the first to differentiate the bass note of a queen about to hatch and the virgin queen's treble piping.

He too rejects many erroneous ideas, including those who say new-hatched bees go immediately to forage – no, only after a few days do they venture around the hive, 'very strictly marking the same' and gradually spanning the whole neighbourhood. They recognise colour, so: 'Forget not to paint the mouths of your hives… that the bees may know their own home.' They can fly a mile in a few minutes and their wings rag out with age. The queen's wings are not short; her body is longer. But her tongue is short, as it is not her job to forage. Uncertain about their life-span, he thinks Moffett's thirty years are a great exaggeration!

Thorley thoroughly enjoys the microscope's revelations too; details that eluded Swammerdam are now visible. He has seen the gut and the honey bag. Such marvellous details adorn God's creatures. Mouths serve as hands, teeth as weapons, antennae as feelers, and feet possess 'a most delightful variety of objects' – sharp claws for grass and sponge pads for smooth surfaces. He notices their glistening velvet dress and the burrowing barbed sting. Of insects in general he rhapsodises:

> *What a profusion of colouring! Azure, green and vermillion; gold, silver, pearls, rubies and diamonds; fringe and embroidery on their bodies, wings, heads and every other part!*[21]

The source of wax was an enduring mystery. From Aristotle to Thorley's time, people thought it was collected directly from flowers or formed from a combination of pollen and venom.[22] Professor Bradley said bees collect wax dust of different colours. Acerbic as ever, Thorley dismisses all such ideas as pure fancy and declares, 'How has [Bradley] betrayed his great ignorance, and abused his Reader?'[23] Wax is always white and never dust. Having seen particles on young bees' bellies, Thorley concludes they make it. In fact a French peasant already knew this but his report to Réaumur was lost. Huber[24] and Burnens observed the fact anew but the actual glands were discovered by Dr John Hunter in 1792.

Elegant and Gruesome Experiments

Equipped with observation hives, professional beekeepers and experimental scientists continue their detailed studies of *apis mellifera* (Linnaeus 1758). Much of Réaumur's major *Histoire des Insectes* is devoted to them, with a particular interest in the hive's marvellous architecture, political structure and swarm formation. Also, according to Cotton (1842), he discovers that a bee's tongue inflates to lick concave surfaces!

20 John Thorley (1744)
Melisselogia, Ch 3 passim.

21 Thorley, p69.

22 Jan Swammerdam (1758)
p161.

23 Thorley, p132.

24 François Huber (1789)
*New Observations on the
Natural History of Bees*, p232.

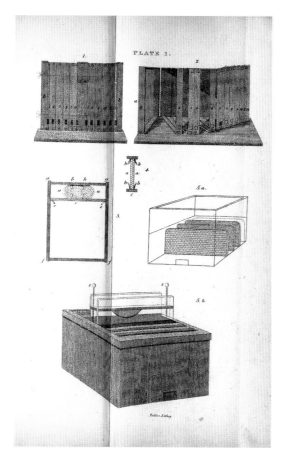

Fig. 17-3 : François Huber's Leaf Hive, from *New Observations on the Natural History of Bees* (1841 ed.) Plate I
© *Oxford University Museum of Natural History*

With their book-like Leaf Hive literally shedding new light on many particulars, Huber and Burnens show a special interest in relations between the queen and the colony. In a chilling experiment they disprove Swammerdam's theory that wingless queens would no longer lay. Even after amputation of all four wings and her antennae, she continues, though without antennae she moves with extraordinary vivacity, dropping eggs carelessly and avoiding her attendants. She eats reluctantly and uncertainly. Robbed of antennae, queens also show no animosity to each other. Huber thus concludes that 'the antennae are not a frivolous ornament… but … organs of touch or smell'.[25]

While continuing to care for their mutilated dowager, workers welcome a stranger queen. Huber further explores the colony's familiar agitation when queenless, ignoring their young and rushing about making a peculiar humming sound. If a new queen's introduction is deliberately delayed, they then eagerly attend her once she starts laying.[26]

Varying such experiments, Warder removed and maimed their queen before spilling a swarm on the grass, with the usual 'piteous and discontented' result. They searched even back to yesterday's resting place but joyfully rushed together upon her return. Poignantly, he observed that they would have stayed with the poor creature until they perished, had he not hived them.[27] When separated by a mesh barrier penetrated only by smell, he finds:

> … an infinite number… inquiring in all directions, and the queen answering these anxious questions… in the most marked manner; for she was fastened by her feet to the grate, crossing her antennae with theirs.

ROYAL BATTLEFIELDS

The scientists analysed conflicts already noted in fanciful terms by Rusden. On introducing a fertile stranger queen, Huber finds sentinel bees threatening to suffocate her on the step. Enraged when he tries to rescue her, their random stinging kills several fellows and the queen. But a colony instantly accepts an interloper after twenty-four motherless hours.

25 Huber (1808 ed.), p261.

26 Huber, pp135-40.

27 Joseph Warder (1722) *The True Amazons*, p46ff.

Réaumur studies rivalrous queens. He has never seen workers kill supernumeraries, as the German Schirach and Riem think they do, but has often watched queens fighting. Within ten minutes of hatching, a young queen tore apart rivals' cells and stung mature occupants to death. The workers then dragged away the remains. When two queens emerged simultaneously:

> *… they rushed together, apparently with great fury, and were in such a position that the antennae of each was seized by the teeth of the other; the head, breast and belly [juxtaposed]; the extremity of their bodies had only to be curved that they might be reciprocally pierced with the stings and both fall dead together. But it seems as if nature has not ordained that…*[28]

…for then they parted precipitately, apparently because the workers could not unite for one party. Had they joined the fray, both queens might have died. When battle resumed, agitated workers restrained them for over a minute but finally one queen gained the advantage and delivered the fatal sting.

'DIVERS MANNERS AND DIFFERENT MOTIONS BY WHICH THEY *UNDERSTAND* ONE ANOTHER'

In noting these manners, Wildman observes that:

> *… when a bee that is at work on the combs demands honey of one that is just arrived; she that wants the honey extends her proboscis and puts it between the talons of her that is to give the honey; in proportion as the latter disgorges it, the other receives it with the proboscis, without spilling a drop.*[29]

The bees rest by hanging onto each other by their feet in long heaps, which must weigh heavily but, he drily notes, they must find this agreeable.

Studies of physiology and behaviour continue, with sometimes petulant argument.[30, 31] In the nineteenth century, Dr Edward Bevan[32] adds details about brood feeding and bee senses. They can hear but with what organs? Two anecdotes convince him they react to tone rather than their master's words; one beekeeper quells angry bees by exclaiming 'Get along, you little fools!' and the other with 'Ah! Would you dare!'

28 Huber, p121-2.

29 Thomas Wildman (1768) *Treatise on the Management of Bees*, p30.

30 John Keys (1796) *The Ancient Bee-Master's Farewell* refers to 'warm disputes… between different naturalists and apiarian societies on the continent' and his own failure to replicate Mr Schirach's 'new and wonderful principles' (p.xii-ix).

31 Robert Huish (1844 ed.) *Bees: their Natural History and General Management* asserts that Huber deserves knocking from his unjustified eminence on the 'pinnacle of apiarian science'.

32 Edward Bevan (1838) *The Honey Bee; its Natural History, Physiology and Management*, p188.

Fig. 17-4 : Title page of Bevan's *The Honey Bee*, 1838
© *Oxford University Museum of Natural History*

Karl von Frisch[33] studied many aspects besides famously decoding the bee dance. He proves that their extraordinarily intricate antennae provide their acute sense of smell and function as a stereo guide to the source. Further exquisite experiments demonstrate connections between antennae, brain hemispheres and memory functions. They use polarised light for navigation but can only distinguish four colours – blue/purple, blue/green, yellows and ultraviolet. Scarlet poppies and white flowers appear ultraviolet. A reward system taught them to differentiate simple shapes but subsequent scientists have proved extreme visual speed is crucial to their foraging among wind-ruffled flowers.

DANCING BEES

Von Frisch's mapping of the insects' dance language is fundamentally important but the behaviour was already familiar. Without fathoming why, Aristotle records how, after foraging, 'they shake themselves and three or four follow'.[34] John Evelyn had also noticed how other bees immediately copy the scouts' shivering motion.[35] The German Reverend Ernst Spitzner describes the signals in 1788:

33 Karl von Frisch (1927, transl.1954) *The Dancing Bees.*

34 J.B.S. Haldane (1955) 'Aristotle's Account of Bees' Dances', in *Journal of Hellenic Studies* LXXV.

35 E. Crane (1999) *The World History of Beekeeping and Honey Hunting*, p567.

Fig. 17-5 : The dancing bee's shimmy is evident in the centre of the picture
Courtesy of Paul Embden

*Full of joy, she twirls in circles about those in the hive, from above downwards and
from below upwards, so that they shall surely notice the smell of honey on her; for
many of them soon follow when she goes out once again.*[36]

Bees dancing to discuss swarming was noticed by Wildman in 1768 and has been
followed up since.

Fig. 17-6 : Karl Von
Frisch's drawing of the
bees' round dance
© *Bayerische Staatsbibliothek
München, Germany*

Von Frisch, however, analyses the foragers' different Round, Waggle and Tremble
dances, considering 'the drama… worthy of the pen of one of those great classical
poets'. This is his vivid description.

*The foraging bee, having got rid of her load, begins to perform a kind of 'round
dance'. On the part of the comb where she is sitting she starts whirling around in a
narrow circle, constantly changing her direction, turning now right, now left,
dancing clockwise and anti-clockwise in quick succession, describing between one
and two circles in each direction. The dance is performed among the thickest bustle
of the hive. What makes it so particularly striking and attractive is the way it infects
the surrounding bees; those sitting next to the dancer start tripping after her, always
trying to keep their outstretched feelers in close contact with the tip of her abdomen.
They take part in each of her manoeuvrings so that the dancer herself, in her madly
wheeling movements, appears to carry behind her a perpetual comet's tail of bees.*[37]

Many have been unwilling to credit bees with such sophistication but later studies
confirm Von Frisch's interpretation of the language.

C.G. Butler[38] of Rothamsted Research says the dance is inborn but the language
takes several days to perfect. Setting bees off from different starting points, he proved
they are not following scouts or scent cues but complying with instructions; the given
bearings cause the poor bees to miss the goal. More recently that experiment has
been elaborated at Rothamsted and in Greenwich and Germany; radar transponders[39]
have confirmed their bearers' obedience to the code, which Riley finds 'pretty
convincing'. Further fascinating findings from Rothamsted are that Asian bees have a
different dance but can decipher and follow European signals!

36 Quoted by J.L. and
C.G. Gould (1988)
The Honey Bee, p55.

37 Karl von Frisch (1927;
1954 transl. Dora Isle and
Norman Walker) *The Dancing
Bees*, p114-5.

38 C.G. Butler (1954)
The World of the Honeybee.

39 J.R. Riley et al,'Recent
Applications Of Radar To
Entomology' in *Outlooks
on Pest Management* –
February 2007, pp1-7.

OTHER BEWITCHING TWENTIETH CENTURY CONTRIBUTIONS

Annie Betts (1884-1961) was perhaps the first woman to make a name for herself in apian research. As a science and engineering graduate applying her First World War aeronautical research to bees' flight, she must certainly have been rare. Brother Adam, renowned for scientific bee breeding, also discovered an innate clock that resists jet lag and enables them to forage appropriate plants at appropriate times wherever they have landed.

Rothamsted's Butler studied how pheromone signals fix social roles within the colony and also refined our understanding of their sensory perceptions. Sharing our four basic tastes: sweet, sour, salt and bitter, they are registered through 'peg organs' on antennae and feet as well as mouths. Without conventional hearing equipment, they are nevertheless highly sensitive to their own sounds. For instance, dancing bees will shift frequency to stand out from the background hum.

Fig. 17-7 : Radar bee
© *Rothamsted Research Ltd*

NEW FRONTIERS

Many new aspects of bee behaviour arise, written up in scientific journals. A superficially zany project at Sydney's Macquarie University discovers that bees on cocaine[40] behave like humans. They dance enthusiastically to communicate honey finds but their performance crashes when undergoing 'cold turkey'. Identifying the responsible genes could possibly lead to improved addiction treatments.

Finally, a close look at pollen, that minute but fundamental life force. Like pebbles on a pretty beach, the varied hues are obvious in this much enlarged photograph of individual grains. With only a paint brush and traditional microscopes, Dorothy Hodges had researched and beautifully painted these cargoes.[41] She would surely have been delighted and astonished by the variety of patterns and intricate layers in each grain now revealed by electron microscopes. 🐝

40 Reported in *The Guardian* on 23rd December 2008, from the *Journal of Experimental Biology.*

41 Dorothy Hodges (1952) *The Pollen Loads of the Honeybee: a guide to their identification by colour and form.*

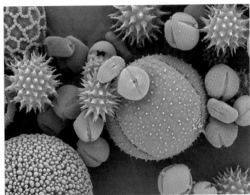

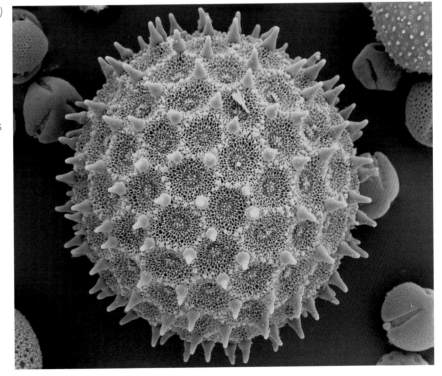

(Clockwise from above)

Fig. 17-8 :

Pollen grains

Courtesy of Paul Embden

Fig. 17-9 :

Mixed pollen grains under an electron microscope

Fig. 17-10 :

Morning glory pollen grain under an electron microscope

Figs 17-9 and 17-10 courtesy of Louisa Howard, Dartmouth College, New Hampshire, USA

18 : Beekeeping — a Craft and a Delight

From the late sixteenth century onwards, learned and literate private individuals evidently took up the craft of beekeeping again. Apart from hands-on 'noble citizen' beekeepers like Virgil, Roman farmers and landowners generally delegated the job to a slave *mellarius*. Then no doubt it was the lay brothers who looked after monastic apiaries. In feudal England, a bee-ceorl was a humble freeman, whilst in mainland Europe the activities of honey hunters were regulated by their lords.

We learn a little of secular beekeeping during Elizabeth I's reign from Holinshed's 'universal Cosmographie [and] histories of every knowne nation', a book which suffered various misfortunes. Commissioned by the queen's printer, Wolfe, but unfinished when he died twenty-five years later, Holinshed then subcontracted 'Britain and England' to William Harrison. The latter, with some frustrations, 'scambled up… this foule frizeled Treatise' but claims 'an especiall eye to the truth of things'. Regarding bees, he observes that 'we cherish none in trees' but insists there are now plenty of hives:

> … *in manner everywhere that in some uplandish towns there are 100 and 200 of them, although… not so huge as those of the east country, but far less, and not able to contain above one bushel of corn or five pecks at the most.* [1]

In contrast, Poland's hives were:

> … *so great, and combs so abundant, that huge boars, overturning and falling into them, are drowned in the honey before they can recover and find the means to come out.*

'So They Will Love Thee, and Know Thee from All Others'

Newly empirical treatises for practical beekeepers multiply, critically amending customary wisdom. Charles Butler's ground-breaking *The Feminine Monarchie* (from which the heading above is taken) emerges from his parsonage apiary. But two truths have never changed: first, that bees depend on the seasons and second, that they are not domesticated creatures. The beekeeper works on their terms, and the need for a calm demeanour must account for the craft featuring in such redemptive novels as Gene Stratton Porter's *The Keeper of the Bees* (1925), about a First World War veteran, and Sue Monk Kidd's *The Secret Life of Bees* (2002), centred on an unloved child and later filmed.

Butler first discusses how to handle bees with the least chance of being stung. The best times are morning or warm evening while the bees are still out. Hot noon is definitely a bad time. If you must disturb them, wear a veil and gloves, though they will accept an 'unhairy hand… unless much offended'. Leather gloves, hair and

1 Wm Harrison *Elizabethan England*, in Holinshed (1586) *Chronicles of England*. (ed. F.J. Furnivall, 1876) pp175-6.

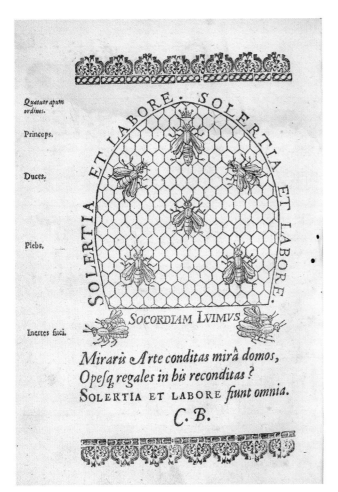

Fig. 18-1 : Frontispiece of Charles Butler's *The Feminine Monarchie*, 1609. Shelfmark Vet A2 e.22 (5), Frontispiece

© *The Bodleian Libraries, The University of Oxford*

feathers provoke them and fustian is an 'offensive excrement'! An angry hive can endanger brainless cattle and even killed a horse innocently nodding off flies and looking over a hedge.

To have their favour, he says you must not be unchaste, uncleanly, sweaty, or stinking of 'onions, Garleeke and the like … Which a cup of beer cures'. Neither be drunk, puffing and blowing, hastily stirring among them or violent in self-defence. Like incorrigible shrews, bees require patience but are reluctant stingers. Softly brush them away, but wash a sting with spittle and 'best be packing as fast as you can' before the smell attracts others.

The Ordering of Bees was published by an admiring son as 'the late unparalleled experience of John Levett, Gent'.[2] Over-protesting his modesty, he asks his Dear Reader whether the world has room for yet more books. But he presses on, bent on correcting past errors. Among numerous florid introductory puffs, one admirer heartily commends:

> *… a work so exact and compendiously done, so plaine and easie for the meanest understanding, yet with all so succinct, deepe, and elaborate, that as a Flie may easily wade in it, so an Elephant may with difficultie swimme in it…*

2 John Levett (1634)
The Ordering of Bees.

Hoping to suit simple beekeepers and not turgid scholars, he avoids airy words, amazing inventions and the intricate windings of a wild brain, he says. However, I have tried without success to imagine a flies' paddling pool that endangers elephants!

The book is addressed to:

> THE VERTUOUS GENTLEWOMAN Mrs Dorothy Kemp, wife to the Right
> Worshipfull Mr Robert Kemp, Esquire, one of His Majesties Justices of the Peace in
> the County of Northfolk…

Perhaps they are the benign patrons of his intended audience:

> … unlearned and Country people, especially good women, who commonly in this
> country take most care and regard of this kinde of commodity, (although much the
> worse for the poor Bees)…

Ouch!

In dialogue form like some Roman manuals, Tortona instructs Petralba, the reluctant legatee of ten hives. His comments are often sharp. Banging basins to catch a swarm is a ridiculous toy. Southern's elaborate method of cleaning an unsavoury hive (described in the quotation below) is unnecessary, and he could never have looked at a comb to think drones are the masters. Levett correctly insists that a hive needs a number of drones but the workers know when to kill them, teaching us 'a remedy against the unmeasurable multitude of our Lawyers'! Supporting unproductive colonies beyond Bartholomewtide (late August) is a foolish custom and a fond conceit. God gave us dominion over more important creatures, so of course we should cull these silly creatures as best benefits us, 'which I take to be the right use of them and the very end of creation'.[3]

WIDENING HORIZONS?

In 1657, Samuel Purchas's *Theatre of Flying Insects* yet again repeats familiar ideas and rejects fond suppositions, but tales from the classics and dubious exotic fragments reappear and he wrongly dismisses a modern observation of the honey bladder's function. Among his sartorial demands, he mentions bees' dislike of beards and long hair. He favours over-wintering viable colonies.

Moses Rusden's final edition of *A Full Discovery of Bees* appeared in 1685, another summation of bee craft. After extravagantly dedicating it to his employer, King Charles II,[4] there follows 'a humble address to the Royal Society at Gresham Colledge'. Though no politician or philosopher, he considers bees better models than Solomon's pismire (ant).[5] Like a responsible foreign correspondent, he will show Nature and Truth naked, stripped of Fancy and ridiculous superstitions; for instance, bees only die after their master if his successor does not know how to care for them. At odds with Levett, he advocates Southern's recipe for hive cleaning:

3 Levett, p41.

4 Reproduced in Chapter 6.

5 Book of Proverbs 6.6. – 'Go to the ant, thou sluggard; consider her ways and be wise…'

6 Moses Rusden (1685), p91.

> … of all the wayes to make the bees best pleased… a hog musling in an hive hath
> been the best way of dressing it, which may be done by throwing in a few handfuls of
> peas, and let an hogg eat them out.[6]

BURGEONING COMMERCIAL INTEREST

In the mid-seventeenth century, national self-sufficiency concerns Hartlib; honey as well as orchards and cider should all be economic crops. We should prefer mead to our enemies' wine, says Warder in *The True Amazons*,[7] his title as unequivocal as Butler's. Better bee management would 'exceedingly help the Poor and delight the Rich', so this 1712 manual claims to spurn 'whimsies and antiquated impertinences' more successfully than, whatever he claimed, Rusden had done.

By the time John Gedde's contribution[8] appears in 1675, mercantile interest is evidently exploding. His small ad in the London *Post Boy* seems to be the first of many by bee chandlers. Besides offering improved hives and recipes for mead, he boasted:

> *This Day is Publish'd, approv'd by the Royal Society,*
> **** The English APIARY, Or, the Complete BEE–MASTER: Unfolding the whole*
> *Art and Mystery of the Management of Bees, Being a Collection and Improvement*
> *of what has been written by all Authors, relating to this Subject, as well Ancient as*
> *Modern...*[9]

In France, too, now that sugar has displaced honey, in 1744 Gilles Bazin sees the possibility of 'very considerable commerce' in wax but business success depends on understanding the insects well. He therefore offers a digest of modern craft and science for ordinary beekeepers, less weighty and detailed than Swammerdam's – or Maraldi, who:

> *... like the gods of the poets, sometimes quit heaven to amuse themselves with*
> *terrestrial creatures; M. Maraldi, I say, diverted himself, amidst his astronomical*
> *observations, with the study of Bees...*[10]

His book is another dialogue. Clarissa recoils from a very sexist Eugenio's rigorous logic, preferring the Ancients' fancies and declaring herself too busy to learn.

> *What would the world say to see the mistress of a family, at the head of a country*
> *farm, go alternately from the examination of a problem, to a review of her poultry,*
> *or from reckoning with her farmers, to a geometrical calculation? Methinks I*
> *would have the awkward grace of those country ladies, who having been at Paris, or*
> *Versailles, mix the air of the court with their own country jargon... I only desire to*
> *know the life, manners, inclinations, the employments, labour and industry of these*
> *little people.*[11]

Pompous Eugenio declares her worthy to be a mother, by which he means 'a great deal. Very few women merit that praise... '. In fact, her unexpected sallies easily disconcert him; on hearing of the queen's sole job, she says, 'Ho! Ho! ... a throne is a dear purchase in this country'. A glass hive and a magnified view of a drone's eye soon engage her enthusiasm and she decides to improve her peasants' skills.

Another quaint throwback to a Virgilian manner is the Jesuit Jacques Vanière's *The Bees: a Poem*[12] to edify a young lady. With all the benefits of Réaumur's researches and Maraldi's glass hives he describes, for instance, the returning forager who:

7 Joseph Warder (1712) *The True Amazons or The Monarchy of the Bees*, pp 28-9.

8 John Gedde (1675) *A New Discovery of an Excellent Method of Bee-houses and Colonies...*

9 19th August 1721.

10 Gilles Bazin (1744) *The Natural History of Bees*, Advertisement.

11 Bazin, p2.

12 Part of a multi-volume work on husbandry, translated by Arthur Murphy (1799).

... grasps the roof, and there suspended clings...
The house-hold train unpack the load with care,
And with the spoil to instant work repair...
But knead and labour for the public use
The ductile wax, and the ambrosial juice;
Or else, like artists in proportion skill'd,
Rooms for the rising generation build...
Neatness prevails; all avenues are clear,
And taste and order through the hive appear...[13]

While tramping well-trodden ground, the Thorleys[14] also want beekeeping to be a national asset. Young Thorley has his own bee chandlery and concludes his book with an invitation to view in their shop many English and foreign honeys and 'all sorts of hives, glass, wood or straw, ready for use'. He also gives twice weekly demonstrations at their Ball's Pond apiary, near Newington Green Turnpike.

In 1766, the new Society[15] for the Promotion of Arts, Manufactures and Commerce launched a financial incentive to commercial apiaries producing a certain quantity of honey and ten pounds weight of merchantable wax. To promote bees' unique value to the rural economy, they also pressed Mills to publish yet another practical guide.[16] He draws attention to 'ignorant' Egypt's migratory beekeeping, taking hives gradually up the Nile to Upper Egypt as seasonal floods recede and flowers bud; like moving sheep to high pastures, this extends their foraging season.

Snippets of bee poetry, equipment small ads and syndicated items in newspapers all indicate growing public interest. The foundation of the Exeter Apiarian Society[17] in 1798 makes national news.

The 'Never Kill a Bee' Movement

Keith Thomas[18] argues that seventeenth century scientific advances blurred the boundary between man and animal more than theologians liked and brought a dawning respect and concern for their welfare. In 1616 an agricultural writer asserts that the bee has 'a kind of wisdom coming near unto the understanding of man' and in 1664 Margaret Cavendish, Duchess of Newcastle, suggests animals might, as fitting to their environment, equal humans in mental and emotional capacity; possibly bees know more than we of flower juices.[19]

As early as 1634, Moffett deplored the destruction, but poisoning bees over burning sulphur remained the norm when harvesting from skeps. However, gaining the honey without destroying the producers is an idea gaining interest. It is one of the Exeter society's objectives and a succession of books advocate the happy convergence of economic efficiency and a less mediaeval view of human rights over Creation.

Settling a swarm in a hive only costs sixpence and two pennyworth of honey but will shortly be worth five to fifteen shillings, says Thorley. Moreover, Gedde's hive, a most happy invention, improves overall management. 'A righteous Man regards the Life of his Beast',[20] and he passionately condemns ignorant mishandling of:

13 Murphy/Vanière, p9.

14 John Thorley (1744) *Melisselogia*, with his son collaborating on later editions.

15 Since 1847, the Royal Society, commonly known as the RSA.

16 John Mills (1766) *An Essay on the Management of Bees.*

17 Founded by Rev Jacob Isaac (1750-1818), Unitarian Minister in Moretonhampstead, Devon.

18 Keith Thomas (1983) *Man and the Natural World: Changing Attitudes in England 1500-1800.*

19 M.L. Cavendish (1664) *Philosophical Letters.*

20 Proverbs 12.10.

Fig. 18-2 : Nineteenth century cottage garden bee shelter, Beeton (1874)

> *... these little creatures whose wonderful parts and properties are so many evident proofs of the infinite power and wisdom of the Creator.* [21]

It is:

> *... notorious Injustice and inexcusable Ingratitude of their cruel and merciless Owners, who not content with all their Treasure (collected with infinite Pains and many Perils) devote them to the Destruction without any Distinction or the least Pity and Compassion. Not unlike so many bloody Ruffians and Murtherers...* [22]

However, he is totally unsentimental in comparing angry, merciless bees to an enraged lion or bear and his instructions on drugging them are fascinatingly precise. A head-sized ball of 'bunt, puckfist or frogcheese'[23] is compressed in a sealed paper bag and dried overnight in a cooling oven. Next day, dry enough to smoulder, an egg-sized piece in a cleft stick is put in an empty hive upended in a bucket to smoke under an occupied hive. In about a minute, the insects will delight you by dropping like hail. When the very last remnant has been shaken onto a table, the queen can often be found among them. Thus briefly doped, small colonies can be blended. Patience, care and a sprinkling of sugar and ale will overcome their natural hostility to each other.

Thomas Wildman[24] was more than the colourful salesman described in Chapter Fifteen; he studied the French scientists and carefully observed his bees. He too wants poor cottagers to keep bees with skill and commends French investment in the craft, particularly a tax rebate, which was given to beekeepers by the city

21 Thorley (1744, 1774) *Melisselogia*, p.xxi-xxii.

22 Thorley, p197.

23 A puffball.

24 Thomas Wildman (1768) *Treatise on the Management of Bees.*

authorities of Rouen. Slaughtering one's colonies is as inhuman and impolitic as killing any other stock for their eggs or milk, he says – but isn't there a logical flaw there?

Members of The Society of Arts joined in, both for economic reasons and through tenderness of heart. Mills points out that no ancient or modern fumes kill every bee, so a few precautions should safeguard all. But more simply, he says contemporary Greeks' suspended combs can be simply lifted while the bees are out foraging, with no smoke tainting the honey. However, if they are to survive the winter, they must be left enough stores. John Keys[25] adds his threnody – 'Alas! Ill-fated bees! Doomed to be victims of your own industry!' He begs cottagers to be humanitarian, even to love their bees.

Other voices dismiss such tenderness and concentrate on monetary returns. Sydserff[26] reckons that burning is more profitable than saving them. William Cobbett[27] addresses his little book *Cottage Economy* (1821) 'To the Labouring Classes of this Kingdom'. This spans from brewing to hat-making, pertinently insisting that 'economy' properly means management, not 'parsimony, stinginess or niggardliness'. Believing they would die of old age before the next harvest, it is whimsical to winter feed a summer swarm. However, bees are very profitable and:

> *He must be a stupid countryman indeed who cannot make a bee-hive; and a lazy one indeed if he will not.*

During the century's political ferment and rapid industrialisation, Cobbett is not alone in pressing good advice on reluctant rustics. Thomas Nutt fell in love with bees as a convalescent boy and published *Humanity to Honeybees* in 1832. In the customary obsequious introduction he finds 'a sort of analogical propriety', given its leading feature, in dedicating it to Queen Adelaide. Humane grounds impress society's higher ranks but conserving the insects profits every beekeeper. His hive design avoids driving them into a new skep while taking their honey, with the forlorn expectation they will restock with brood and winter food. That is:

> *… as painful… as for an industrious man and his thriving family to be rudely ejected from a comfortable house and home, without the least notice… to take shelter in an empty house, and left there destitute…[28]*

Close on Nutt's heels comes Bagster, lamenting the lower classes' 'truly gothic ignorance' and obstinacy:

> *… people averse to all instruction in the management of their bees; their fathers, grandfathers and so on up to Noah, followed this or that method, and therefore it is good.[29]*

That same year, in steps a gentleman student of Christ Church College, William Cotton. His Oxford Apiarian Society aims both to promote better bee management by lending hives and giving prizes to cottagers; and to promote scientific knowledge among the higher classes. He urges:

> *… clergy, gentry and their good wives to… aid me as a united body, in teaching their poor neighbours the best way of keeping bees.[30]*

25 John Keys (1789) *The Practical Bee Master* – one of several works bearing the same name.

26 Robert Sydserff (1792), *Treatise on Bees*.

27 Perhaps better known for his *Rural Rides*.

28 Thomas Nutt (1832; 2nd ed.1834): *Humanity to Honey-bees*, p106.

29 Samuel Bagster (1834) *The Management of Bees, with a Description of the Ladies' Safety Hive*, pp49-50.

30 William Cotton, *Letters to Cottagers*, later incorporated in *My Bee Book* (1842).

Lower class morals are evidently as important as their financial profit. By now, rural allotments excite 'and that most justly, almost universal attention' among the well-meaning,[31] but they involve thirsty work and beer-shops often stand hard by, whereas bee-gardens are only second to a Sunday School. Bees demand chaste and clean behaviour, as Butler said, and keep a man at home, tending, watching and loving them. Finally, 'NEVER kill your bees'. If you burn them, you deserve to be sulphured yourself.

Surely a less draconian course is Earl Spencer's proposal of a Chair in Beekeeping, celebrated in this anonymous ditty.[32]

> *How doth the little bee*
> *Increase her little power,*
> *And gather favour every day,*
> *And almost every hour!*
>
> *In Exhibition hive and tent*
> *She would be sharer too,*
> *So swarms to Kensington are sent*
> *Their busy work to do.*
>
> *And if Earl Spencer should reply,*
> *'The bees shall have a Chair,'*
> *To Kensington again she'll fly…*

Modern Beekeeping

After nearly two centuries of inquisitive practice and a century of increasing commercialisation, the middle of the nineteenth century saw significant technological advances. Besides a settled principle of hive design, innovations included Abbé Collin's queen excluder, to prevent her laying eggs among the honey stores, and De Hruschka's centrifugal spinner, to draw honey from the combs. Maeterlinck noted how only a few years had massively expanded the honey industry. Now man is the bees' master, 'although furtively, without their knowledge…'[33]

Local Beekeepers' Associations spread across the country and the national BBKA, founded in 1874, furthers good practice and supports research. Declining bee populations have stimulated more people to keep them; membership in 2013 stood at over 23,000 and hive numbers have at least doubled since 2007. The summer population of honeybees was estimated in 2012 to be 48 billion but adverse weather since has led to many losses.

Hobby beekeepers have recently brought two divergent philosophies to the craft; one uses a small, high-tech hive that simplifies management and harvesting and the other is the 'barefoot beekeeper' approach, where bees pollinate and produce honey for their own needs with minimal interference. It will be interesting to see whether these ideas make significant incursions on the familiar hive pattern.

31 Nutt, p282.

32 Proposed during an RHS Garden Exhibition in South Kensington. I have been unable to trace whether this materialised, but certainly such Chairs do now exist.

33 Maurice Maeterlinck (1901) The Life of the Bee, p13.

Fig. 18-3 : A contemporary beekeeper inspecting a frame

Courtesy Rufus Reade, photographer, and of Charles Ramsden

THE LEISURED BEEKEEPER

As commercial interest grew, so did beekeeping as a hobby among prosperous people. In 1806, physician John Evans, claiming to know more than Virgil and regretting a dearth of science amid too much poetic moralising, published *The Bees: a Poem*. Originally meant only for his own family, friends persuaded him to share his allusions, ruminations and science with a wider audience.

> *Inspire my lay! While boon Pomona's bowl*
> *Kindles fresh rapture in the Poet's soul,*
> *While echoing songs from Wrekin's brow recoil,*
> *And lives in verse the Farmer's changeful toil,*
> *Oh! Shame to Britain! Shall her sylvan Muse*
> *Still to the Bees their well-earned meed refuse...*
>
> *Some gleams indeed the British Muses yield*
> *Of partial light o'er this neglected field,*
> *Witness great Shakespear!*

Embroidering in Virgilian mode, he chronicles everything from the spring awakening, ending abruptly with the drones' slaughter. A typical passage compares the maiden queen's flight to glory on the Creçy battlefield. (Perhaps deservedly, the book was still uncut in the Bodleian Library when I requested it!)

'THE KEEPING OF BEES IS LIKE THE DIRECTION OF SUNBEAMS' – *Henry Thoreau*

Besides the chatty, sentimental and jokey books quoted in Chapter 15, there are more contemplative works, such as Maeterlinck's classic[34] already often cited.

Feeling nostalgic as modern hive farms displaced picturesque skeps in cottagers' old bee-gardens, a Sussex sage, Edward Edwardes, presented another neo-Virgilian work[35] and *Lore of the Honeybee*. Extending well-trodden trails, he calls on ancient Bardic traditions of this Land of Honey. It might have been Phoenicians mining tin in the West Country 3000 years ago who introduced wild honey harvesting to Albion[36] but probably the Romans taught us beekeeping. Fancifully, he suggests the worker bee's sting became venomous to avenge its uselessness as an egg channel. Suddenly impatient of theorising, he quotes a seventeenth century work, *Country Housewife's Garden*:

> *I leane not on conjectures, but love to set down that I know to be true, and leave these things to them that love to divine.*

After the First World War, an Essex country rector[37] finds solace in bees' dreamy whirring as he walks from garden to church. He wants people to love them; Nature is both our earthly and spiritual habitat. Saint Bernard's honeyed heaven, the hymn's line 'Jerusalem… with milk and honey blest' join others from Rupert Brooke and Horace to comfort him.

> *We have found safety in things undying —*
> *The winds of morning, tears of men and mirth,*
> *The deep night, and birds singing and clouds flying.*
>
> *Oh, may that spot of all the globe be mine!*
> *Hymettus' self not purer honey yields…*

More modern works inform while walking the same contemplative path. In *The Joys of Beekeeping*, Richard Taylor claims the honeybee as his 'good daemon' and regrets manuals that omit delight; he rejoices as they cascade home through the trees. Winston[38] too ranges widely and reflectively around the bee yard and modern science. For instance, he enjoys intricate studies of bee brains, their remarkable cleverness in knowing their lifetime sequence of duties and intelligent foraging strategies, but equally valid is his own pleasure in honey's stickiness and bees walking over his hands. Another American, Bonnie Pearson,[39] speaks of being in love with her bees, slowing down, sitting by the hive, entering into the colony's intimacy. Chandler[40] insists on his 'barefoot' refusal to exploit them. 'Put nothing harmful in. Take only what they can spare.' He stresses literally 'listening to them', interpreting their tones like the old writers did.

Are all beekeepers quiet, solitary contemplatives? Industrialised honey farming in the USA involves ruthless mechanical handling of bees on their combs as well as their long distance mass migration.[41] Small scale beekeepers, however, are calm among the seething mass of an open hive and speak quietly, while shaking down combs with surprising vigour to get a sight of the honey crop and brood.

34 M. Maeterlinck (1901) *The Life of the Bee.*

35 Edward Edwardes (1908) *The Bee Master of Warnlow.*

36 Albion was Eilanbaun in the Celtic language, meaning White Island.

37 Herbert Brown (1923) *A Bee Melody.*

38 Mark L. Winston (1998) *From where I sit.*

39 Teacher of beekeeping in Ohio. See websites.

40 P.J. Chandler (2009) *The Barefoot Beekeeper; a simple sustainable approach…*

41 *More than Honey* (2012) Swiss documentary film by Markus Imhoof.

We have noted many caustic exchanges in centuries of controversy, not least Gedde labelled an ignoramus propagating 'falsehoods and ridiculous directions'. At a recent meeting one quiet man remarked that it was usual for twelve members to have thirteen opinions. In 1951, reading bee papers disabused one reader of the *British Bee Journal* of previous notions:

I had always imagined that these beveiled gentlemen, moving rather ponderously among their hives, were big-hearted, patient and perhaps rather placid...

Countless poets since Virgil have written of bees' wisdom, ambrosial delights and the buzz of sunny afternoons but a few contemporary poet beekeepers have taken very varied standpoints. Sylvia Plath[42] appears profoundly ambivalent, contemplating her God-like power but also appalled by 'unintelligible syllables... like a Roman mob'. Perhaps 'If I stand very still they'll think I'm cow-parsley'. Diana Hartog's fanciful prose poem[43] imagines bees taking flight with her in a honey-filled tub. In a tiny verse welcoming a contented retirement, István Kemény[44] contrasts a hundred years of being an electrician with 6000 years of beekeeping.

42 Sylvia Plath 'The Arrival of the Bee Box' and 'The Bee Meeting'.

43 From Diana Hartog (1992) *Polite to Bees: a Bestiary.*

44 Istvan Kemény (1998) *The Beekeeper.*

THE BEEKEEPER'S YEAR

Spring
Snowdrops crack frozen ground,
Dip their heads under wintry sun. Hives stir —
Over-laden entrails seek relief. They must eat.
A few antennae reach out, retreat.

Fitful sun tempts again; crocus faces shine.
The washing line's aflap and sheets reveal
Telltale stains. Bees grab nectar where they can
For the queen's on her rounds of vacant cots.

Fig. 18-6 :

Spring forager

Courtesy of Paul Embden

A mild March day — you briefly check.
A laying queen? Are brood cells ringed
With well-filled stores? Budding labourers
Must have syrup while blooms are few.

This colony is weak, lacks a lusty queen.
Bring a young pretender with her ladies
Of the bedchamber, safe in a candied cage
Until doubting subjects accept the new regime.

In April all their needs are clamant.
More brood space must be built. Mouths
Multiply apace. Beware this hungry month
Before dandelions sunburst everywhere.

Fig. 18-7 : Summer foraging
Courtesy of Paul Embden

Summer
In high summer each colony multiplies —
As brood breaks out in multitudes
The queen recharges empty cells.
Workers toil through every sunlit hour

In beanfield, garden and a few
Surviving meadows, 'unimproved',
Embroidered with rattle, vetch, birds' eye,
Loading bright pollen panniers and nectar sacs.

Enlarge the hive and clip the royal wings
Or watch for whirling tumult in the air
And lose your swarms unless
You have new hives prepared.

August sees foraging past its peak.
Confident of a young and fertile queen
And readying for winter, they massacre
The drones and cap their stores.

Honey harvest time — smoke them down, take
Your share of laden combs. Slice off the caps
And spin ambrosial gold. Gleaming jars await
Summer's pure and filtered essences.

Fig. 18-8 :
Unwanted squatter

Courtesy of Paul Embden

Autumn
Populations fall and space must shrink
To match. Late flowers and sunny days
Will find them on the wing, but be sure
To give a sufficient syrup store

In gratitude. Help the guards
By narrowing their door against
Sundry unwelcome lodgers
And the season's hungry robbers.

Fig. 18-9 : Sleeping hives
Courtesy of Paul Embden

Winter

Your balled up colonies rest
In the heart of the hive. Toast
Your toes on the hearth and feast
Your tongue with a honey roast.

BIBLIOGRAPHY

Abercromby, Hon. J. 'The Beliefs and Religious Ceremonies of the Mordvins', in *Folklore Journal* vol.vii, pp65-135, 1889.

Adam, Brother *In Search of the Best Strains of Bees*, Northern Bee Books, Hebden Bridge 1983.

Adam, Brother *Beekeeping at Buckfast Abbey*, 4th ed., Northern Bee Books, Hebden Bridge 1987.

Advancis Medical Limited www. medicalhoney.com

Aeneas the Tactician (c.4 BC) *How to Survive under Siege* (transl. Illinois Greek Library), Heinemann, London, 1923.

Aeschylus, transl. Lewis Campbell, *The Persians*, Oxford University Press, London 1906.

Alcock, Mary *Poems* (1799) 'The Hive of Bees; a Fable written in December 1792'.

Allen, W.F., C.P. Ware and L.McK. Garrison *Slave Songs of the United States*, A. Simpson, New York 1867.

Ambrose, John T. 'Insects in Warfare', in *Army*, 33-8, 1974.

Anon *Letters on Entomology: intended for the Instruction and Amusement of Young Persons...*, printed for Geo. B. Whittaker, 1825.

Apollonius Rhodius, transl. Edward P. Coleridge, *Argonautica*, G.Bell, London 1889.

Appian of Alexandria, transl. J.D. Denniston, *Roman History*, Book 12 'Mithridatic Wars', Harvard University Press, London, 1912-13.

Aristotle, transl. A.L. Peck, *Generation of Animals*, Heinemann, London 1943.

Aristotle, transl. A.L. Peck, *History of Animals* Vol. II, Heinemann, London 1970.

Aristotle, ed. and transl. D.M. Balme, *History of Animals* Books vii-x, Harvard University Press, London 1991.

Auden, W.H. & Taylor, Paul B. (transl.) *The Elder Edda: A Selection*, Faber, London 1969.

Augustine, *Saint Of the Good of Marriage*.

Austen, Ralph (1653) *The Spirituall Use of an Orchard*, Printer Theo. Robinson, Oxford 1847.

Baden-Powell, Robert *Scouting for Boys*, C. Arthur Pearson Ltd, London 1908.

Bagster, Samuel *The Management of Bees, with a Description of the Ladies' Safety Hive*, Samuel Bagster, or William Pickering, London 1834.

Baker, Henry *Of Microscopes and the Discoveries Made Thereby*, J. Doddesley, London 1742, new ed. 1785.

Baker, Richard A. and Wit Chimielski 'How old are bees? A look at the fossil record', in *Journal of Apicultural Science* 47.1, pp.79-86, 2003.

Baring, Anne and Jules Cashford *The Myth of the Goddess: Evolution of an Image*, Viking, London 1991.

Bartolomaeus Anglicus (1240) *On the Properties of Things*

Bartolomaeus Anglicus (ed. Robert Steele) *Mediaeval Lore from Bartolomaeus Anglicus*, Alexander Moring Ltd, London 1905.

Basil, Saint *Hexaemeron*, Sixth Homily on the Days of Creation, Part 4.

Basil, Saint 'Letters', transl. Sr A.C. Wray, *Fathers of the Church*, vol. 13. Catholic University of America Press, Washington DC 1951.

Bazin, Gilles Augustin (1744) *The Natural History of Bees*, transl. from the French.

Beach, F.A. 'Beasts Before the Bar' in Alan Ternes (ed.) *Ants, Indians, and Little Dinosaurs*, Charles Scribner's Sons, New York 1975.

Bede, Venerable (7th century, transl. Leo Sherley-Price 1955) *History of the*

English Church and People, Penguin, Harmondsworth.

Bell, Gary M. 'Gilpin, George (d. 1602)', *Oxford Dictionary of National Biography*, Oxford University Press, Oxford 2004.

Benjamin, Alison and Brian McCallum *A World without Bees*, Guardian Publications, London 2008.

Bérubé, Conrad *War and Bees: Military Applications of Apiculture*. www3.telus. net/Conrad/war-bees.htm (2000).

Bevan, Edward *The Honey Bee; its Natural History, Physiology and Management*, Van Voorst, London 1838.

Bevington, Louisa S. ed. *Liberty Lyrics*, Liberty Press, London 1895.

Boyd, Doug *Rolling Thunder – a personal exploration into the secret healing powers of an American Indian medicine man*, Random House, New York 1974.

Bracegirdle, Brian (ed.) *Beads of Glass: Leeuwenhoek and the Early Microscope: catalogue of exhibition* at Science Museum, London, and Museum Boerhaave, Netherlands, 1982.

Brackenbury, John *Insects and Flowers: a Biological Partnership*, Blandford Press, London 1995.

Bradbury, S. *The Evolution of the Microscope*, Pergamon Press, Oxford 1967.

Bricker, Victoria R. and Helga-Maria Miram (transl and annotated) *An Encounter of Two Worlds: The Book of Chilam Balam of Kaua*, Tulane University: Middle American Research Institute, New Orleans, 2002.

Bromenshenk, Jerry J. *et al* 'Can Honeybees Assist in Area Reduction and Landmine Detection?' in *Journal of Mine Detection* VII/3, 380-9, 2003.

Brooks, E.W. *Joseph and Asenath*, SPCK, London 1918.

Brown, Herbert *A Bee Melody*, Andrew

Melrose Ltd, London 1923.

Brown, R. *Beekeeping: a seasonal guide*, Batsford, London 1985.

Brown, Ron *Great Masters of Beekeeping*, Bee Books New and Old, Somerset 1994.

Butler, Charles *The Feminine Monarchie*, 1609 and 1623 editions.

Butler, C.G. *The World of the Honeybee*, Collins, London 1954.

Campbell, Patrick (1793) *Travels in the Interior inhabited parts of North America in the years 1791 and 1792*, Chaplain Society, Toronto 1937.

Campion, Alan *Bees at the Bottom of the Garden*, A&C Black, London 1984.

Cantimpré, Thomas de *The Mediaeval Bestiary*, 13th century.

Castaldo, Stefano and Francesco Capasso, 'Propolis, an old remedy used in modern medicine', *Fitoterapia*, Volume 73, Supplement 1, November 2002, pages S1-S6. Elsevier *Science Direct* 2010.

Chandler, P.J. *The Barefoot Beekeeper: a simple sustainable approach to small scale beekeeping using top bar hives* 3rd ed. 2009. www.biobees.com

Chapiro, Jacques, *La Ruche*, Flammarion, Paris 1960.

Charles-Edwards, T. and F. Kelly (ed. and transl.) *Bechbretha: an old Irish Law-tract on Beekeeping*, Dublin Institute for Advanced Studies, Dublin 1983.

Chaucer, Geoffrey *The Canterbury Tales*, 14th century.

Cheshire, Frank R. *Bees and Beekeeping: Scientific and Practical*, L. Upcott Gill, London 1886.

Christenson, Allen J. *Popol Vuh: Sacred Book of the Quiché Maya People*, www.mesoweb.com

Clark, J.F.M. *Bugs and the Victorians*, Yale University Press, London 2009.

Cobbett, William *Cottage Economy*, Harris Edwards Publications, Shrewsbury 1821.

Collier, Mary *The Woman's Labour*, 1739.

Columella, Lucius Junius Moderatus (fl. c.1 AD) *De Re Rustica* (transl. Forster, E.S. and Edward H. Heffner) *On Agriculture* Book IX, Heinemann, London 1954.

Combe, Wm *Dr Syntax's Three Tours: in search of the Picturesque, Consolation, and a Wife*, Chatto & Windus, London 1895 ed.

Cook, A.B. 'The bee in Greek mythology' in *Journal of the Hellenic Studies* 15. 1ff, 1895.

Cooper, Diana *Bee Tidbits*, Tufts University, Medford, Mass. USA, 1985. www.chebucto.ns.ca/~ag151

Cotton, W.C. *My Bee Book*, J.G.F. & J. Rivington, London 1842.

Cotton, W.C. *Buzz a Buzz or The Bees, done freely into English... from the German of Wilhelm Busch, Griffith and Farron*, London 1872.

Cowper, William *Complete Poetical Works of William Cowper*, Oxford University Press, Oxford 1907.

Crabbe, George *The Parish Register: Burials*, 1807.

Crane, E. *Honey: a Comprehensive Survey*, Heinemann and Bee Research Association, London 1976 ed.

Crane, E. *The Archaeology of Beekeeping*, Duckworth, London 1983.

Crane, E. *The World History of Beekeeping and Honey Hunting*, Duckworth, London 1999.

Crane, E. *Making a Beeline*, IBRA, 2003.

Crane, E. 'Retrospect: use of bees in warfare', in *Bee World* 84.2, p94-7, 2003.

Croft, L.R. *Curiosities of Beekeeping*, Northern Bee Books, Hebden Bridge 1990.

CSIRO Australia (4th March 1999), 'Bees – Latest Weapon In Cancer Fight'.

Cuff, John *A Description of the Most Valuable Kinds of Microscopes now in use: viz. the Pocket Microscope together with the New Invention for fixing it on a Pedestal and giving Light to Objects by a Speculum*, London 1758?

Dahl, Roald 'Royal Jelly' in *The Best of Roald Dahl*, Michael Joseph, London 1983.

Dante, *Paradiso*, Canto XXXI, transl. Henry Francis Cary (1844) Folio Society, London 2009.

Day, John *The Parliament of Bees* Benj. Harris, London 1641.

Dexter, Miriam Robbins, and Karlene Jones-Bley (eds) 'The fall and transformation of Old Europe: Recapitulation 1993' pp. 351-372 in *The Kurgan Culture and the Indo-Europeanization of Europe: Selected articles from 1952 to 1953 by Marija Gimbutas*, Washington DC: Institute for the Study of Man, 1997.

Dickens, Charles *Bleak House*, Oxford World Classics Edition 2008.

Diodorus Siculus *Historical Library*, transl. C.H. Oldfather, Heinemann, London 1933-67.

Dobson, Jessie 'Some Eighteenth Century Experiments in Embalming', in *Journal of the History of Medicine and Allied Science*, VIII, October 1953.

Dunning, J.M. *The Key of the Hive*, Kingsgate Press, London 1945.

Edwards, David B. 'Mad Mullahs and Englishmen: Discourse in the Colonial Encounter' in *Comparative Studies in Society and History* 31.4, pp 649-70, October 1999.

Edwards, Lin 'Bees Helping to Monitor Air Quality at German Airports', on PhysOrg.com, 21st July 2010.

Edwardes, Edward Tickner *The Lore of the Honeybee*, Methuen, London 1908; reprint 1920.

Edwardes, Edward Tickner *The Beemaster of Warrilow*, Methuen, London 1920.

Ehrenberg, Margaret *Women in Prehistory*, British Museum Press, London 1989.

Encyclopaedia Judaica, 2nd ed., eds. Skolnik, Fred, Berenbaum, Michael, 1945-2007 || Detroit : Macmillan Reference USA in association with the Keter Publishing House.

Engel, Michael S., Hinojosa-Diaz, Ismael A., Rasnitsyn, Alexandr P. 'The Biogeography of Apis (Hymenoptera; Apidae;Apini)', in *Proceedings of the California Academy of Sciences*, Series 4, 60, 3, pp.23-38, 2009.

Erler, Mary Carpenter, and Maryanne Kowaleski (eds.) *Gendering the Master Narrative*, Cornell University Press, London 2003.

Euripides *The Bacchae*, transl. Philip Vellacott. Penguin, Harmondsworth 1953.

Euripides *Iphigenia in Tauris*, transl. Maurice Platnauer, Clarendon Press, Oxford 1938.

Evans, Jeremy, and Sheila Berrett *The Complete Guide to Beekeeping*, Unwin Hyman Ltd, London 1989.

Evans, John *The Bees: a Poem*, Shrewsbury 1806.

Evelyn, John (17th century) *Diaries*, Clarendon Press, Oxford.

Fife, A.E. *The Concept of the Sacredness of Bees, Honey and Wax in Christian Popular Tradition*, University of Utah: PhD thesis 1939.

F.P., or John Gage *The Christian Sodality, or Catholick hive of bees sucking the hony of the Churches prayers from the blossomes of the Word of God, blown out of the Epistles and Gospels of the Divine Service throughout the year*, 1652.

Fraser, H.M. *Beekeeping in Antiquity*, University of London Press, London 1931.

Frankum, Richard *The Bee and the Wasp: a Fable in Verse*, with etchings by George Cruikshank, London 1832, 2nd ed. 1861.

Frith, Jacquelyn, Ruth Appleby, Rebecca Stacey and Carl Heron 'Sweetness and Light: Chemical Evidence of Beeswax and Tallow Candles at Fountains Abbey, North Yorkshire', in ads.ahds.ac.uk *Notes and News* vol. 48. 220-227.

Gay, John *The Works of Mr John Gay in 4 vols* vol.3. 'The Degenerate Bees' London. 1769; 1772 ed.

Gayre, G.R. *Wassail! in Mazers of Mead*, Phillimore & Co., London 1948, 1986 ed.

Gedde, J. *A New Discovery of an Excellent Method of Bee-houses and Colonies*, D. Newman, London 1675; 3rd ed. 1677.

Gibran, Kahlil *The Prophet*, 1926.

Gimbutas, Marija *The Goddesses and Gods of Old Europe, 6500-3500 BC.* Thames and Hudson, London 1982.

Giuman, Marco *Melissa: Archaeologia delle Api e del Miele nella Grecia*, Giorgio Bretschneider Editore, Rome 2008.

Goodison, Lucy, and Christine Morris *Ancient Goddesses: The Myths and the Evidence*, British Museum Press, London 1998.

Gordon, R.K. *Anglo-Saxon Poetry*, J.M. Dent & Sons, London 1942.

Gould, J.L. and C.G. Gould *The Honey Bee*, Scientific American Library, New York 1998.

Grimaldi, David A. and Michael S. Engel *Evolution of the Insects*, Cambridge University Press, Cambridge 2005.

Grout, Roy A. (ed.) *The Hive and the Honeybee*, Dadant & Co., Hamilton, Illinois 1946.

Haldane, J.B.S. 'Aristotle's Account of Bees' Dances', in *Journal of Hellenic Studies* LXXV, 24-5, 1955.

Hall, Joseph *Occasional Meditations*, London 1630.

Harrison, Jane Ellen *Prolegomena to the Study of Greek Religion*, Cambridge University Press, Cambridge, 3rd ed 1922.

Harrison, Wm (1587; ed. Lothrop Withington) 'Elizabethan England' in Holinshed's *Chronicles of England*, Walter Scott, London 1902.

Hartlib, Samuel *The Reformed Commonwealth of Bees*, 1655.

Hartog, Diana *Polite to Bees: a Bestiary*, Coach House Press, Toronto 1992.

Harvey, Paul *A Medieval Oxfordshire Village: Cuxham, 1240 to 1400*, Oxford University Press, London 1965.

Harvey, William *Anatomical Exertations Concerning the Generation of the Animals*, 1651.

Henderson, William *Notes on the Folklore of the Northern Counties of England and the Borders*, Longmans, Green & Co., London 1866.

Herbeman, Charles (ed.) 'Early Christian Lamps', in the *Catholic Encyclopaedia* 1913.

Herodian *Roman History*, transl C.R. Whittaker, Book 3.9.38. Heinemann, London 1969.

Hesiod (7th century BC) *Works and Days*, transl. David W. Tandy and Walter C. Neale, University of California Press, London c.1996.

Hesiod *Theogony*, transl. G.W. Most, Harvard University Press, London 2006.

Higgins, Reynold *Greek and Roman Jewellery*, Methuen, London, 2nd ed 1980.

Hodges, Dorothy *The Pollen Loads of the Honeybee: a Guide to their Identification by Colour and Form*, Bee Research Association, London 1952.

Hill, Selima *Gloria: Selected Poems*, Bloodaxe Books, Northumberland 2008.

Homer *The Iliad* (transl. R. Lattimore) University of Chicago Press, Chicago 1951.

Homer *The Odyssey* (transl. R. Lattimore) Harper Collins, New York 1965.

Hooke, Robert (1665) *Micrographia, or Some Physiological Descriptions of Minute Bodies, Made by Magnifying Glasses with Observations and Inquiries*. Facsimile edition, Dover Publications, New York 1961.

Horn, Tammy *Bees in America: How the Honey Bee Shaped a Nation*, University Press of Kentucky, Lexington 2005.

Huber, François (1789) *New Observations on the Natural History of*

Bees, Thomas Tegg, London, 2nd ed 1808; 1841 ed.

Hudson-Williams, T. 'King bees and queen bees' in *Classical Review* 49, 2-3, 1935.

Huish, Robert, FZA *Bees: their Natural History and General Management: Comprising Full and Experimental Examination of the Various Systems of Native and Foreign Apiarians; with an Analytical Exposition of the Theory of Huber; Containing also the Latest Discoveries and Improvements in Every Department of the Apiary*, Henry G. Bohn, London 1844.

Isack, H.A. (1999) 'The role of culture, traditions and local knowledge in co-operative honey-hunting between man and honeyguide: A case study…', in Adams, N.J. & Slotow, R.H. (eds) *Proc. 22 Int. Ornithol. Congr., Durban: 1351-1357.* Johannesburg: BirdLife South Africa.

Isocrates to Demonicus 52 (transl. George Norlin). Heinemann, London 1928.

Jardine, William (1840) *The Naturalist's Library*, vol. 6, p.1955, Wikimedia Commons.

Jefferson's *Notes on the State of Virginia 1785*, Electronic Text Center, University of Virginia Library.

Jones, Richard *The Eva Crane Historical Collection: The Folk Art of Slovenian Hive Fronts*, Rhondda Cynon Taf, IBRA, 2013.

Johnson, James W. 'That Neo-Classical Bee', *Journal of the History of Ideas* 22.2, pp. 262-266, 1961.

Jones, Ray *Bee-Sieged: Bees in Warfare*, Barny Books, Grantham 2008.

Jonstonus, Johannes (1631) *A History of the Wonderful Things of Nature*, translated into English by a Person of Quality, London 1657 ed.

Josephus, transl. H. St J. Thackeray, *Jewish Antiquities*, Heinemann, London 1958.

Kearney, Ray *St Augustine's Marriage and Virginity*, New City Press, New York 1999.

Kelly, Fergus, and T.M. Charles-Edwards (1983 ed.) *Bechbretha: an Old Irish law-tract on bee-keeping*, Dublin Institute for Advanced Studies, Dublin 1983.

Kelly, Kevin *Out of Control: The New Biology of Machines, Social Systems, and the Economic World*, www.kk.org 2003.

Kerenyi, Karl (transl. Ralph Manheim) *Dionysus: Archetypal Image of Indestructible Life*, Routledge Keegan Paul, London 1976.

Keys, John *The Ancient Bee-Master's Farewell; or Full and Plain Directions for the Management of Bees to the Greatest Advantage*, G&G Robinson. London 1796.

Keys, John *The Practical Bee-Master: in which will be shewn how to manage bees either in straw hives or in boxes, without destroying them and with more ease, safety and profit than by any other method made public*, London 1780.

Kezic, Nikola 'Brief Note on Research on Mine Detecting Bees in Croatia' *Journal of ERW and Mine Detection* 13.1., 2009.

Kidd, Sue Monk *The Secret Life of Bees*, Review, London 2002.

Kramer, Samuel Noah *From the Tablets of Sumer: thirty-nine firsts in man's recorded history*, Falcon's Wing Press, Indian Hills, Colorado 1956.

Laroia N. and D. Sharma. 'The religious and cultural bases for breastfeeding practices among the Hindus', in *Breastfeeding Medicine*, 1, 94-98, 2006.

Larrington, Carolyne (transl.) *The Poetic Edda*, Oxford University Press, Oxford 1996.

Larrington, Carolyne *The Feminist Companion to Mythology*, Pandora, London 1992.

Levett, John *The Ordering of Bees, or The True History of Managing Them*, T. Harper, London 1634.

Locker, Frederick *Patchwork*. Smith, Elder & Co., London 1879.

Lockwood, Jeffrey A. *Six-Legged Soldiers: Using Insects as Weapons of War*, Oxford University Press, Oxford 2008.

Longgood, Wm F. *The Queen Must Die, and Other Affairs of Bees and Men*, Norton, London 1985.

Lönnrot, Elias (transl. Keith Bosley) *The Kalevala*, Oxford University Press, Oxford 1989.

Mackail, J.W. *Select Epigrams from the Greek Anthology*

Maeterlinck, Maurice (1901; transl. A Sutro) *The Life of the Bee*, Allen, London 1909.

Mandeville, B. de *The Fable of the Bees, or Private Vices, Publick Benefits*, J. Rorerts, London 1714.

Mandevillle, Sir John *Travels* (1900 Macmillan edition). www.bestiary.ca/ beasts/beast1195.htm

Mangold, Tom and John Penycate *The Tunnels of Cu Chi*, Pan Books, London 1985.

Marnix, Philips van (1569; transl. George Gilpin) *The Beehive of the Romishe Churche. A worke of al good Catholikes too bee read and most necessary to bee understood: wherein both the Catholike religion is substantially confirmed, and the heretikes finely fetcht ouer the coales.* T. Dawson, London 1579.

Mayor, Adrienne *Greek Fire, Poison Arrows and Scorpion Bombs: Biological and Chemical Warfare in the Ancient World*, Duckworth, London 2003.

Medibee Limited www.medibee.co.uk

Merrick, Jeffrey *Order and Disorder under the Ancien Régime*, Cambridge Scholars Publishing, Newcastle 2007.

Mills, John *An Essay on the Management of Bees*, J. Johnson and B. Davenport, London 1766.

Milne, A.A. *Winnie-the-Pooh*, Methuen, London 1926.

Moffett/Mouffet, Thos *Insectorum*, or *The Theater of Insects, or, Lesser Living Creatures*, London 1634/1658

Montaigne, Michel de *Essays 1580-88*.

Morse, R. and T. Hooper *The Illustrated Encyclopaedia of Beekeeping*, Blandford Press, Poole 1985.

Mosley, Leonard *Duel for Kilimanjaro: The East African Campaign 1914-18*, Weidenfeld & Nicholson, London 1963.

Muldoon, Paul *Collected Poems 1968-1998*, Faber and Faber, London 2001.

Murphy, Arthur (transl.) *The Bees: a Poem from the Fourteenth Book of Vanière's Praedium Rusticum*, F. and C. Rivington, London 1799.

Nakajima, Y. *et al* 'Comparison of bee products based on assays of antioxidant capacities' in *BMC Complementary and Alternative Medicine*, 9.4. 2009.

Neufeld, Ed. 'Insects as Warfare Agents in the Ancient Near East', in *Orientalia* 49, pp.30-57, 1980.

Newbolt, Henry (ed.) *The Book of Cupid, an Anthology*. London 1909.

Nutt, Thomas *Humanity to Honey-bees: or, practical directions for the management of honey-bees upon an improved and humane plan; by which the lives of bees may be preserved, and abundance of honey of a superior quality may be obtained*, Longman & Co., London 1832; 2nd ed. 1834.

Orchard, Andy *Cassell's Dictionary of Norse Myth and Legend*, Cassell, London 1997.

Ovid *Fasti* (transl. Sir J.G. Frazer), Heinemann, London 1989.

Paley, William 'The Fable of the Beehive', in *Reasons for Contentment Addressed to the Labouring Part of the British Public*, London 1831.

Palladius (4th century BC) *De Re Rustica*, Book 1: Governaunce 19. Middle English translation by Mark Liddell, E. Ebering, Berlin 1896.

Pausanias (2nd century AD, transl. W.H.S. Jones) *Description of Greece*, Heinemann, London 1935.

Pepys, Samuel *Diaries 1665* Fullbooks. com

Pliny *The Natural History* Book XI (transl. H. Rackham) and Book XXI (transl. W.H.S. Jones) Heinemann, London 1947 and 1951.

Plutarch *Lives*, 'Solon', 2nd Century AD.

Poinar, G.O., Jnr, and B.N. Danforth 'A Fossil Bee from Early Cretaceous Burmese Amber', in *Science* vol.314, no. 5799, 2006.

Pomeroy, Sarah B. *Goddesses, Whores, Wives and Slaves*, Pimlico, London 1975.

Porter, Gene Stratton *The Keeper of the Bees*, Doubleday, Page & Co, New York 1925.

Potts, Daniel T. *Mesopotamian Civilization: The Material Foundations*, Cornell University Press, Ithaca, New York 1997.

Preston, Claire *Bee* Reaktion Books, London 2006.

Preston, Mollie *Diary of a Farmer's Wife*, Countrywise Books, London 1964.

Puckle, Bertram S. *Funeral Customs*, T. Werner Lawrie, London 1926.

Purchas, Samuel *A Theatre of Politicall Flying Insects*, Thomas Parkhurst, London 1657.

Raleigh, Sir Walter *History of the World*, 1614.

Ramirez, Juan Antonio *The Beehive Metaphor: from Gaudi to Le Corbusier*, Reaktion Books, London 2000.

Ransome, Hilda M. *The Sacred Bee in ancient times and folklore*, George Allen & Unwin, London 1937.

Réaumur, René-Antoine Ferchault de *Notes to serve for a history of insects*, 1741.

Paul Rehak 'The Isopata Ring and the Question of Narrative in Neopalatial Glyptic', from: kuscholarworks.ku.edu/ dspace/bitstream/1808/8364/1/ Rehak_Isopata.pdf

Riley, J.R., J.W. Chapman, D.R. Reynolds and A.D. Smith 'Recent Applications of Radar To Entomology' in *Outlooks on Pest Management*, pp1-7, February 2007.

Rolle, Richard 'The Nature of the Bee', in Kenneth Sisam (ed.) *Fourteenth Century Verse and Prose*, Oxford University Press, London 1937.

Root, A.I. (1908, 2nd ed. E.R. Root) *The ABC and XYZ of Bee Culture*, Medina, Ohio 1948.

Rothamsted Research (Bee Dept.) 'Waggle dance controversy resolved by radar records of bee flight paths', press release 12th May 2005.

Ruestow, Edward G. 'Piety and the Defense of the Natural Order: Swammerdam on Generation', (p217-41) in *Religion, Science and World View: Essays...*, ed. Margaret Oslers and Paul L. Faber. Cambridge University Press, New York 1985.

Rusden, Moses *A Full Discovery of Bees*, London 1685.

Sackville West, Vita *All Passion Spent*, Hogarth Press, London 1931.

Saxo Grammaticus *The History of the Danes* (transl. Oliver Elton), Norroena Society, New York 1905.

Schacker, Michael *A Spring without Bees: how Colony Collapse Disorder has endangered our food supply*, Lyons Press, Guildford, Connecticut 2008.

Scheinberg, Susan 'The Bee Maidens of the Homeric Hymn to Hermes' in *Harvard Studies in Classical Philology* 83, pp 1-28. 1979.

Scott Moncrieff, C. K. (transl. 1921) *Beowulf*, The Westminster Press, London.

Seneca ad Lucilium *Epistulae Morales* (transl. Richard M. Gummere), Heinemann, London 1943.

Seneca De Clementia *Of Mercy I*, in *Morals* (transl. John W. Basore), Heinemann, London 1958.

Shakespeare, William *Henry IV, Part 1* and *Henry V*.

Simon, Jean-Baptiste, *Le Gouvernement Admirable*, 1740.

Skilling, Robert N.H. *Sixty Years with Smoker and Veil* Northern Bee Books, Mytholmroyd 1991.

Smart, C. *The Poetical Works of Christopher Smart* (ed. Karina Williamson), Clarendon Press, Oxford 1987 and 1996.

Smith, D.A. 'Notes and Comments on John Evelyn's Manuscript on Bees from Elysium Britannicum' in *Bee World* 46.4, pp116-131, 1965.

Sophocles *Oedipus at Colonus* (transl. Robert Fagles) Penguin, Harmondsworth 1984.

Spooner, David *Thoreau's Vision of Insects*, USA:X-Libris.com 2002.

Steiner, Rudolf (transl. Thomas Braatz) *Bees*, Anthroposophic Press, Hudson, New York 1998.

Stiles, Kristine 'Uncorrupted Joy: International Art Actions', p321-2 in Schimmel, P. '*Out of Actions: Between Performance and the Object, 1949-79*', Thames & Hudson, London 1998.

Strabo, *Geographia* Book IV (1st century BC, transl. H.L. Jones) Heinemann, London 1969.

Sturluson, Snorri (13th century; transl. Henry Addams Bellows) *Poetic Edda*, American-Scandinavian Foundation, New York 1923.

Sturluson, Snorri (transl. L.M. Hollander) *Heimskringla*, University of Texas Press, Austin 1964.

Swammerdam, Jan *The Book of Nature*, C.G. Seyffert, London 1758.

Taylor, P.B. and W.H. Auden *The Elder Edda* (transl.), Faber, London 1969.

Taylor, Richard *The Joys of Beekeeping*, Barrie & Jenkins, London 1976.

Tautz, Jürgen *The Buzz about Bees*, Heidelberg:Springer-Verlag, Berlin 2008.

Thomas, Keith *Man and the Natural World: Changing Attitudes in England 1500-1800*, Penguin, Harmondsworth 1984.

Thorley, John *Melisselogia, or the Female Monarchy*, London, 1st ed. 1744; 4th 1774.

Topsell, Edward *The History of Serpents or The Second Book of Living Creatures*, London 1658.

Varro, Marcus Terentius (1st century BC) *De Re Rustica* (transl. Wm Davis Hooper, revised by Harrison Boyd Ash), Heinemann, London 1934.

Virgil *Georgics IV* (30 BC) transl. Peter Fallon Oxford University Press, Oxford 2006; transl. J.B. Greenough (19th century) Tufts University online; transl. John Dryden, World's Classics, London 1903.

Virgil *Eclogues* (37 BC, transl. H. Rushton Fairclough) Harvard University Press, London 1916.

Virgil *The Aeneid* (19 BC, transl. W.R. Jackson-Knight) Penguin, Harmondsworth 1956.

Von Frisch, Karl (1927; transl. Dora Isle and Norman Walker) *The Dancing Bee*, Methuen, London 1954.

Von Frisch, Karl *You and Life* (transl. Ernest Fellner and Betty Inskip), John Gifford Ltd, London 1940.

Walker, P. and Richard Jones (eds) *Eva Crane, Bee Scientist, 1912-2007*, IBRA, Cardiff 2008.

Warder, Joseph (1712) *The True Amazons or The Monarchy of Bees*, London 1722, 1726 eds.

Westenholz, Joan G. 'Goddesses of the Ancient Near East 3000-1000 BC', pp63-81, in Goodison and Morris 1998.

White, Gilbert (1788) *The Illustrated Natural History of Selborne*, Webb & Bower, Exeter 1981.

Wildman, Thomas *Treatise on the Management of Bees*, T. Cadell, London 1768.

Wilson, Bee *The Hive*, John Murray, London 2004.

Winslow, Pat *Kissing Bones*, Templar Poetry, Matlock 2012.

Winston, Mark L *From Where I Sit: Essays on Bees, Beekeeping and Science…* Cornell University Press, Ithaca, New York 1998.

Wither, George *The Schollers Purgatory*, London 1624.

Wodehouse, P.G. *The Drones Omnibus*, Hutchinson, London 1982.

Xenophon (4th century BC) *Anabasis*, Book IV, transl. E. D. Stone, Macmillan, London 1912.

Xenophon *Economics VII*, transl. E.C. Marchant, Heinemann, London 1923. Xenophon *Hellenica* (trans. Rex Warner as *A History of My Times*), Penguin, Harmondsworth 1966.

Miscellaneous sources

Burney Collection of Seventeenth/ Eighteenth Century newspapers, Bodleian Library, Oxford.

Nineteenth/Twentieth Century newspapers on microfilm, Bodleian Library, Oxford.

Beecraft (miscellaneous copies).

British Bee Journal November 1891, 29th July 1920, 19th August 1950.

The Scottish Beekeeper, March 1974.

Angelopoulos, Theodorus *The Beekeeper* film fiction 1986.

Butler, Martin and Bentley Dean, narrated by Ernie Dingo, *First Footprint*, Australian television documentary on ABC, October 2013.

Simmons, Jeremy '*The Last of the Honey Bees*', television documentary on More 4 13th October 2009.

Markus Imhof (director) *More than Honey*, documentary film 2013.

www.ancientsites.com/aw/ Article/1170885#

www.andrewgough.co.uk

www.iap.gr Archaeological Receipts Fund Directorate of Publications, Athens.

www.mothergoddess.com

www.phrases.org.uk Gary Martin, *The Phrase Finder*.

www.scholastic.com/browse/article. jsp?id=3754880 'Biodetective Bees, 13th September 2010'.

www.thebeegoddess.com

www.ucmp.berkeley.edu/history/ leeuwenhoek.html

INDEX